Minerals, Collecting, and Value across the U.S.-Mexico Border

Minerals, Collecting, and Value across the U.S.-Mexico Border

ELIZABETH EMMA FERRY

INDIANA UNIVERSITY PRESS

Bloomington and Indianapolis

This book is a publication of

INDIANA UNIVERSITY PRESS
Office of Scholarly Publishing
Herman B Wells Library 350
1320 East 10th Street
Bloomington, Indiana 47405 USA

iupress.indiana.edu

Telephone orders   800-842-6796
Fax orders   812-855-7931

© 2013 by Elizabeth Emma Ferry

Manufactured in the United States of America

Library of Congress Cataloging-in-Publication Data

Ferry, Elizabeth Emma.
  Minerals, collecting, and value across the U.S.-Mexico border / Elizabeth Emma Ferry.
      pages cm — (Tracking globalization)
  Includes bibliographical references and index.
  ISBN 978-0-253-00928-9 (cloth : alk. paper) — ISBN 978-0-253-00936-4 (pbk. : alk. paper) — ISBN 978-0-253-00948-7 (electronic book) 1. Minerals—Collection and preservation—United States. 2. Minerals—Collection and preservation—Mexico. 3. Rock collectors. 4. Mineral industries—Social aspects. 5. United States—Commerce—Mexico. 6. Mexico—Commerce—United States. I. Title.
  QE392.5.U5F47 2013
  382'.45549972—dc23
2013001007

1 2 3 4 5   18 17 16 15 14 13

To
David, Sebastian, and Isaiah

CONTENTS

ACKNOWLEDGMENTS

So many people have helped me with this project, it is hard even to know where to begin. However, I will start in Guanajuato, where I first got the idea from seeing miners sell minerals, use them as religious offerings, and give them as gifts. My particular thanks go to Cirilo Palacios and his family, Pancho and Domingo Granados, Alejandra Gómez, Elia Mónica Zárate, and Ada Marina Lara Meza. In Mexico City, thanks to María Guadalupe Villaseñor, Juan Carlos Miranda, Oscar Irazaba, and Oscar Escamilla. Among those whom I met in Tucson and Colorado, I am particularly indebted to Dennis Beals, Peter Megaw, and Wendell Wilson, as well as to Terry Wallace, Steve Smale, Tom Gressman, Mike New, Herb Obodda, Gene Schlepp, and Carole Lee, among others. In Cambridge, Massachusetts, thanks to Carl Francis, Alden Carpenter, and the members of the Boston Mineral Club, especially Jim Catterton and Nate Martin. In Mapimí, Lázaro de Anda and Mario Pecina were particularly kind and helpful. At the Smithsonian Institution, Pamela Henson, Jeffrey Post, Pete Dunn, and James Luhr went out of their way to educate and guide me. Lawrence Conklin, William Panczner, and George Hoke provided me with valuable historical information and materials. Rubén Lechuga Paredes and Vera Regehr, both, at the time, doctoral students at the Universidad Iberoamericana, served as gifted research assistants in Mapimí. Thomas Moore of the *Mineralogical Record* reviewed the geological and mineralogical discussions in the book and went above and beyond in editing my prose.

The project has also benefited from the tremendous help of my anthropological and other colleagues, particularly: Mark Auslander, Manduhai Buyandelger, Josiah Heyman, Sarah Hill, Robert Hunt, Smita Lahiri, Sarah Lamb, Ann Marie Leshkowich, Mandana Limbert, Caitrin Lynch, Roger Magazine, Carlota McAllister, Janet McIntosh, Paul Nadasdy, Richard Parmentier, Heather Paxson, Smitha Radhakrishnan, Leslie Salzinger, Karen Strassler, Ajantha Subramanian, Christine Walley, and David Wood. My thanks go, as always, to my advisors and mentors, especially to Katherine Verdery, Michel-Rolph Trouillot, Sidney Mintz, Gillian Feeley-Harnik, and Fernando Coronil. Many thanks to Robert Foster for his support for this project in an early phase and his patience as I slowly got it finished. Thanks to my editor and assistant editor Rebecca Tolen and Sarah Jacobi for all their help and guidance and to two anonymous reviewers for Indiana University Press. I am also indebted to the Smithsonian Institution Fellowship Program, The Newhouse Center for the Humanities at Wellesley College, and the Brandeis Latin American and Latino Studies Program and Norman Fund for Faculty Research. Audiences at Brandeis University, Cornell University, Harvard University, Wesleyan University, SUNY Albany, Boston College, el Colegio de Michoacán, and Universidad Iberoamericana heard earlier versions of some chapters and provided useful comments. An earlier (and quite different) version of chapter 5 was published in *American Ethnologist* under the title "Geologies of Power: Value Transformations of Minerals from Guanajuato, Mexico."

MINERALS, COLLECTING, AND VALUE
ACROSS THE U.S.-MEXICO BORDER

Introduction:
Making Value and
U.S.-Mexican Space

This book traces the movements of minerals—discrete bits of the earth's crust like the ones commemorated in two series of postage stamps issued in the United States and Mexico (figures 0.1 and 0.2)—as they circulate from Mexican mines through markets and museums in Mexico and the United States. These objects are valued in many different ways: as scientific artifacts, collectibles, religious offerings, commodities (some cheap, some very pricy), and gifts. This book explores the range of things that people in Mexico and the United States think about and do with minerals, as well as what minerals do as actors in their own right. These practices surrounding minerals depend on mining, museum and private collecting, and scientific research, all crucial areas in the relationship between Mexico and the United States over the past 150 years. I look at the transactions through which minerals are created as valuable, and further, at how people and minerals create value together and thus create many other things: objects, knowledge, people, places, markets, and so on. This attention to value gives us a new perspective on the United States and Mexico and the connections between them. But to begin thinking about these bigger questions, we need some idea of what kind of things we are talking about. What do I mean by "minerals?"

Definition: Mineral

> 1. A naturally occurring inorganic element or compound having an orderly internal structure and characteristic chemical composition, crystal form, and physical properties.[1]
>
> —*Dictionary of Mining, Mineral, and Related Terms*

From this definition, we already know several things. Minerals are not made by humans. They are not organic. Because they have an orderly

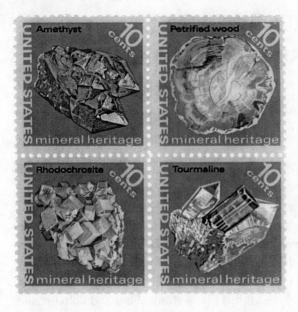

FIGURE O.I. "America's Mineral Heritage" U.S. postage stamps, 1974.

FIGURE 0.2. "Minerales Mexicanos" Mexican postage stamps, 2005.

internal structure, they are not gases or liquids. They are identifiably distinct materials—that is, they are not rocks, which are agglomerations of minerals formed through geologic processes. So far, so good.

However, this only takes us part of the way to understanding the protagonists of this story, which can be defined far more specifically. Minerals can be melted down as ore or cut into gemstones. We ingest them in our food and water and make them into components of objects such as watches, radios, lampshades, and bombs. They can be used in many ways, although most of these instances lie outside the scope of this book. I focus on minerals that are used as distinct objects in their own right rather than as ingredients or components of something else, in the form of by-products of ore mining, scientific specimens, collectors' specimens, religious offerings, and natural art. I am primarily concerned with three fields where minerals are valued: ore mining, mineral collecting, and mineralogy. All of the minerals I consider here are found in Mexico and are used in Mexico and the United States.

A few illustrations may help make clear the kinds of issues and objects under consideration.

Denver, Colorado, 2005: At the Denver Gem and Mineral Show, in one of the hotels where dealers rented rooms to display their wares, I met a middle-aged U.S. man looking at some trays set up near the vending machines. As we peered at thumbnail specimens of malachite and azurite (green and blue copper minerals), I told him about my research. He responded enthusiastically and said, "A mineral person looking at a mineral is like a mother looking at her baby. It's a spiritual thing. Sometimes when a stone is coming to me, I will dream in that color for weeks. There's a deep pleasure there."

Mapimí, Durango, 2007: A dealer who runs the small store at the municipal museum in Mapimí, a dusty mining town in northern Mexico whose population has shrunk over the course of the twentieth century, invited me to his house to see his mineral collection. He told me of his life in minerals. "My father always knew I would be connected to 'el risco' [the mineral business]," he said. "My family took a picture of me as a baby sitting on a table, surrounded by minerals from [the] Ojuela [mine]."

Saturday Evening Post, 1927: George Kunz, mineral collector and gem expert for Tiffany's, gave an interview under the title, "American Travels of a Gem Collector." In it he described his adventures while collect-

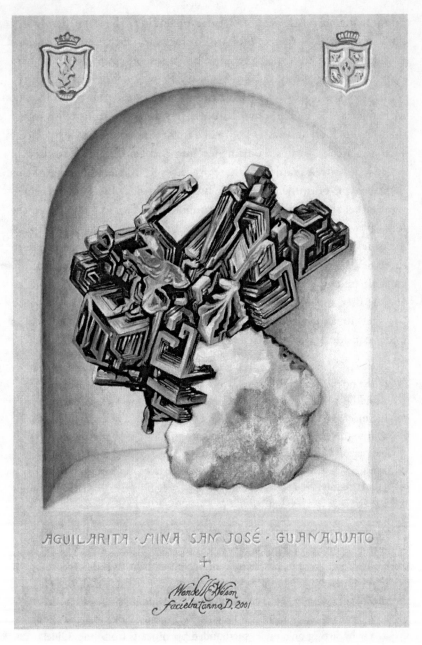

FIGURE 0.3. Aguilarite, Mina San José, Guanajuato. Oil on canvas, 9×12 inches (2001). Painted in the style of the seventeenth-century Spanish still-life painters, from a 4.2-cm cluster of aguilarite crystals from the San José mine, Guanajuato, Mexico, in the Terry Wallace collection. Artist's collection ©Wendell Wilson 2001. Reprinted with permission.

FIGURE 0.4. Roadside altar with minerals. Photo by Elizabeth Ferry.

ing in Mexico: "For the seeker of gems Mexico offers its treasures of jade, obsidian, turquoise and opal. Though a semiprecious stone, the reddish-yellow opal of Mexico—the finest in the world—is worth up to $1500; but as usual, it is not the price but the whole surrounding drama of their formation in Nature, their discovery, the adventure of going out to seek them, their mineralogical nature and significance, and their marketing which constitute their interest for the gem expert" (22–23).

Mapimí, Durango, 2008: My research assistant and I interviewed Felix Esquivel, a mining prospector who found one of the world's most expensive mineral specimens, "The Aztec Sun." In 1977, Esquivel sold it as part of a lot of 25 specimens for around US$4,000; it reportedly sold again recently for US$1.7 million. He said, "They called it 'the big stone,' but it wasn't big. It was in the form of a cross. [It] was on TV and in books, but we don't have anything. Later people came to see if I had any others, but I didn't, and then I couldn't work anymore."

Santa Rosa, Guanajuato, 1998: I visited a miner's house in the small mountain town of Santa Rosa, where many miners live. He showed me a box of minerals he stores until the Christmas season, when he places them near the crèche his family sets up in their home. When I admired the arrangement, he wished out loud that he could make me a gift of one of the minerals, but said, "I can't give these to you, they belong to the Baby Jesus."

Mexico City, early nineteenth century: The Spanish-Mexican mineralogist Andrés Del Río gave classes in mineralogy at the Colegio de Minería in Mexico City. He traveled around Mexico working out technical solutions at mines and smelters, but his first love was mineralogy. On one occasion, he said, "I am more interested in a little piece of some new or curious genus or species the size of a nut than in a rich nugget of gold or various quintales [a quintal was a hundred pounds] of Batopilas silver." (Arnaíz y Freg 1936:29)

Tucson, Arizona, 2009: Asked about the sale of the Aztec Sun legrandite, which he helped to broker, a prominent dealer told me, "I can say that it sold for a price that was pushing 2 million dollars. It was the biggest price for a mineral that wasn't a tourmaline ever. That's what people don't realize—the most beautiful thing in our world is only two million dollars. Compare it to a Van Gogh—what's that, some smears on a canvas. That sells for way more than nature's best creation."

These images, descriptions, and comments show how people's uses and experiences of Mexican minerals are embedded in rich histories of mining, science, devotion, and collecting. These histories go back to the mid-sixteenth century, when silver was first exploited on a large scale in the mines of Zacatecas, Guanajuato, Taxco, and elsewhere. The mining districts of Zacatecas and Guanajuato each took a turn as the world's leading silver producer (in the sixteenth and eighteenth centuries, respectively), and over the years, many other economically valuable metals were found, including gold, copper, zinc, and lead, and scores of mining localities were founded. Dense nodes of economic activity, these mining localities tended to bring together not only miners, but also ranchers, traders, farmers, and many others to serve the complex needs of the mines and their workers.

Mining centers such as Guanajuato and Zacatecas lived and died on the global price of metals, and mining affected not only economic activities in a strict sense but also religion, kinship, and cosmology. A visit to any of Mexico's "silver cities" (e.g., Zacatecas, Guanajuato, Taxco, Real del Catorce, or Mapimí) reveals silver's tidemarks. Churches and public buildings, neighborhoods, elite and popular genealogies, holidays, art and music, cosmology and religious practice, and heritage tourism all can be traced back to silver in one way or another. Minerals, mined substances that occur alongside silver and other metals, show these traces when they appear on altars, in local museums or public buildings, and in people's houses. The miner in Santa Rosa who presented minerals to the Baby Jesus follows a tradition of placing minerals on domestic altars and altars inside the mines. A chapel in the Templo del Señor de Villaseca in the Cata neighborhood of Guanajuato is lined with quartz and amethyst from the nearby mines, and I found tombstones in both Guanajuato and Mapimí that were encrusted with minerals.

The geology of mining centers shows complexity equal to the social formations I just described. Many substances besides ore emerged from Mexico's mines, some of which proved useful to the industry. The study of *mineral paragenesis* (the location and combination of minerals as a result of geological forces) was necessary to plan exploration and production, especially as mining was rationalized in the nineteenth century. These needs on the part of the mining industry provided an impetus for what we would call applied mineralogical and geological research. The rise of the earth sciences in eighteenth-century Europe planted the seeds for geology and mineralogy in the New World, with the Real Seminario de Minería (Royal Mining Seminary), renamed the Colegio de Minería after Mexican Independence, as the vanguard institution. People began to study minerals not only for their immediate practical use in mining, but also to advance the scientific study of the earth.

Mexican mines also frequently produced colorful and intricately formed crystallized minerals that became collectors' items, first in Europe and soon after in the New World. Sometime in the late nineteenth century a trade in mineral specimens emerged in mining centers. Selling minerals provided extra income for miners and a hedge against volatile metals prices (and therefore uncertain wages and employment). As markets developed for minerals, they absorbed part of the working population as small-scale dealers who sold to buyers from Europe and the United States. Meanwhile, opals, amethysts, and other semiprecious minerals came into fashion. The opal mines of Querétaro and amethyst mines of Guerrero and Veracruz became famous, developing markets and attracting foreign visitors. In his *Saturday Evening Post* article, George Kunz described his visit to an opal mine, saying:

They had been working these mines for a century and yet, as we looked up the height of rock, there, peering and winking at us like myriads of curious eyes, shone thousands upon thousands of these bright opals, from lucent pastel to the rich red of the fire opal. They gleamed like little electric lights flashing on and off, as the sunbeams faltered on them, flaming like beast eyes when a beam of light strikes them through the night. There at the mine I went over the hoards of opals, each one a miniature sunset as it lies in your palm, like a shower of fireworks as they pour from your fingers. (1927:23)

In the United States, mineralogy and mineral collecting developed more slowly than in Mexico primarily in scholarly circles and small societies in New Haven, Connecticut; Cambridge, Massachusetts; Philadelphia, Pennsylvania; and New York, New York. After the Civil War, westward expansion and the search for a transcontinental railroad route opened up mining centers in Arizona, Colorado, and California and drew money and infrastructure for the earth sciences. Mineralogical research began to take off and mineral specimens became necessary for advanced research and for training. The Smithsonian Institution, the American Museum of Natural History, Harvard University, and a few other institutions acquired bequests and conducted field research to put together world-class collections. This spurred mineral markets in Mexico, which became progressively more complex.

In the twentieth century, descriptive mineralogy, the mineralogy subfield most dependent on mineral specimens, declined somewhat. On the other hand, the appeal of mineral specimens as collectibles grew, then skyrocketed. Today mineral collecting ranges from field collecting and local mineral clubs, to roadside rock shops and prefabricated sets for young scientists, to scientific collectors in museums and universities, to an elite and glamorous set of collectors who sell and trade minerals for many thousands and sometimes even a million dollars. This is partly due to an analogy now drawn by some collectors and dealers between minerals and fine art. The dealer's exclamation that minerals are "nature's best creation," far better than "some smears on a canvas," exemplifies this attitude.

The gem and mineral shows in Tucson, Arizona; Munich, Germany; Ste.-Marie-Aux-Mines, France; Denver, Colorado; and Costa Mesa, California draw thousands or tens of thousands of visitors. Dozens of dealers sell on the Internet, through mineral auction houses, and through magazines for mineral connoisseurs. In places such as Guanajuato, and Mapimí, Mexico, dealers maintain complex commercial webs and compete fiercely with rivals and interlopers, while miners give minerals away or place them on altars, tombstones, and on the tops of bureaus or curio shelves.

Who inhabits these worlds defined by the circulation of mineral specimens? How do they use and experience minerals? How do the minerals themselves draw people together or separate them? How have the multiple ways they are valued—as scientific specimens, collectibles, devotional objects, commodities, and gifts—changed over time, and how have these changing values affected the worlds that people and minerals inhabit? As we look for answers to these questions, it becomes clear how closely connected the United States and Mexico have been since before their births as independent nations. We are taken beyond familiar tales of economic imperialism and see how the two nations have emerged and grown through both exploitation and mutual exchanges, and how people and things continually create relations between the two countries through action and transaction.

A crucial sphere of action in which people and things create the worlds they inhabit is that of making value. Through attempts to create things as valuable, the social and material world is stabilized in enduring ways, in things such as buildings, institutions, national boundaries, markets, scientific journals, museum collections, churches, altars, and tombs. We must attend carefully to the precise circumstances through which minerals are made valuable to understand how minerals participate in creating the United States and Mexico in relation to one another.

Minerals are a good choice for telling this story, for several reasons. The range of contexts within which these objects acquire value—mines, scientific laboratories, gem and mineral shows, museum exhibits, domestic altars, and collection displays, to name only the most significant— allows us to see how many different ways they can be valued, and by extension, the multiple nature of value-making itself. The case of Mexican minerals in Mexico and the United States also shows how new modes of valuing minerals have emerged and stabilized over the past fifty years. Some of these modes have now come to play a dominant role in the world of mineral collecting, with far-reaching economic and social consequences. I trace how specific actors and actions have helped to crystallize these forms of value, which over time have come to appear as intrinsic, permanent qualities.

Minerals have an abrupt materiality, a "thingness," that makes them particularly apt for a study of the production of value in and through objects. Their physical qualities make certain kinds of relations and interpretations more available than others. That minerals are hard, inorganic, and usually very old makes them seem an especially "material" form of matter, and their consistent patterns of crystallization, luster, and other qualities make them seem particularly stable and timeless. Thus minerals are good to think about how value becomes solidified in the material world. In an essay about olive oil, anthropologist Anne Meneley describes the qualities of the oil as qualisigns, as defined by the philosopher Charles Peirce, or signs that derive their meaning from an intrinsic quality. Meneley enjoins us to "tarry a bit longer with the sensuous materiality of the symbols themselves; after all, one cannot make a potent symbol out of just anything" (2008:308). In this sense, minerals form the material substrate for materiality itself.

The particular qualities of minerals also help us to think about one of anthropology's most vital questions: How do humans create multiple worlds that are grounded in material things and territorialized in places but that are also emergent and mobile? Minerals are the quintessence of grounded, territorialized matter (they are, after all, chunks of place) that also move around in space. As such they provide us with a way of thinking through the dialectics of place and mobility that are central to human experience.

Studying minerals sheds light on aspects of U.S.-Mexican relations that are not often discussed in anthropology. Other ethnographies of the United States and Mexico have focused on migration, electoral politics, and popular culture. Mining, mineralogy, and mineral collecting are areas that may seem immune to sociocultural analysis, but they yield a wealth of information about the two countries and their sociocultural interactions. Studying minerals points up several distinctive aspects of U.S.-Mexican relations: mining and its relation to colonial and postcolonial arrangements and to national sovereignty and state formation; territorial contraction/expansion and the vexed zone of the border; and circulations and disruptions of technoscientific knowledge and expertise.

My study has two main aims. First, I focus on the everyday practices and transactions of people and things to see how U.S.-Mexican transnational space—the spatial relations and experiences conditioned by but not reducible to territorial nation-states—is gradually built up. Second, the study mobilizes the anthropology of value to see exactly how transnational space is built up over time. I see the ongoing creation of value as essential to the constitution of spatial experience and look at practices involving Mexican minerals through the lens of value-making.

These two projects—showing a new dimension of U.S.-Mexican space, and showing how in making value people and objects make the world around them—are intimately connected. The multidimensional transnational space of the United States and Mexico, and the people and objects that inhabit it, are to a great extent the products of, as well as subsequent producers of, successful value-making. This is because value-making creates and organizes difference—and not just any difference, but "meaningful difference" (a term I adapt from the work of David Graeber [2001]). The process of arranging meaningful difference, when it works, brings people and things together in more or less stable configurations, from which new attempts to make value can be launched. I do not use this phrase in an abstract or figurative sense, but in relation to a specific and concrete set of actions and phenomena. In the case of Mexican minerals, many of these configurations are spatial in character, such as marketplaces and routes between them, the paths that people and minerals follow, or the physical sites of mineral shows, scientific expeditions, mines, schools of mines, museums, and so on.

Transnational Space and Technoscience

In framing this study of minerals in terms of the gradual accretion of social-material spaces, my approach owes a debt to Henri Lefebvre's magisterial work *The Production of Space* (1991 [1968]), which introduced the notion of social space as a historical product specific to the mode of production of a given social formation. Lefebvre's insight was that space is not an a priori category or blank stage on which action takes place. Space has to be made through social action. Lefebvre is talking about the space that corresponds to entire social orders, but others have made a similar point in more concrete and local terms. For instance, in his essay "How to Get from Space to Place in a Fairly Short Stretch of Time: Phenomenological Prolegomena" (1996), Edward Casey puts analytic priority on the ways specific places are located through bodily experience. Tim Ingold introduces the concept of the "taskscape" to describe the ways that sites and landscapes are sites constituted through human labor and activity (Ingold 1993) while Don Mitchell's book *The Lie of the Land* (1996) traces the role of migrant workers in shaping the California landscape. Moreover, where Lefebvre makes a distinction between *social space* and *physical space,* at least conceptually, Casey, Ingold, and Mitchell see sociality and materiality as intimately connected, a perspective that has strongly informed my study. Minerals, after all, make up the physical earth, and the transactions between minerals, collections, money, scientific descriptions, photographs, dealers, miners, and collectors between Mexico and the United States are vitally constitutive of spaces that are simultaneously material and social.

A concern with transnational social-material space is part of a now highly elaborated conversation among those who study Mexico (and other places too). The concept of transnationalism was developed by anthropologists, geographers, and others in the 1990s to describe social and material relationships among people and things that regularly crossed over (while still being affected by) national political borders (Basch et al. 1994; Hannerz 1996). The "transnational" aimed at capturing phenomena not well described as either national or international. Its defining feature was hybridity: Transnational phenomena demonstrated features that emerged out of two different national contexts but could not be reduced to either one. The concept of transnationalism also emphasized people's lived practice and experience; it described phenomena not from the perspective of geopolitical institutions, state actors, or important people, but "from the ground up."

Because many transnational phenomena were expressed in terms of social or material space, often deeply bound up with movement and circulation, distance and proximity, *transnational space* became a more specific and, in some instances, more appropriate term than *transnationalism* (Gupta 1992; Crang et al. 2004; Tolentino 1996). Particular national pairings or diaspora became commonly described in terms of transnational space, most prominently the United States and Mexico, but also Germany and Turkey and other places bound together by migration and other forms of circulation, as well as individual nations with intensive diasporic reach, such as Sri Lanka or the Philippines.

In an article published in 1991, "Mexican Migration and the Social Space of Postmodernism," Roger Rouse wrote,

> We live in a confusing world, a world of crisscrossed economies, intersecting systems of meaning, and fragmented identities. Suddenly, the comforting modern imagery of nation-states and national languages, of

coherent communities and consistent subjectivities, of dominant centers and distant margins no longer seems adequate. Certainly, in my own discipline of anthropology, there is a growing sense that our conventional means of representing both the worlds of those we study and the worlds that we ourselves inhabit have been strained beyond their limits by the changes that are taking place around us. (8)

Rouse's argument explores how Mexican migration to the United States overturns the classic "sociospatial images" that shaped scholars' views of rural Mexico up until that time (and since) (9). The article stands as an excellent early formulation of how dramatic increases in Mexican migration to the United States and efforts to understand these increases and their effects helped to create new concepts and methodologies in the discipline. Rouse addresses two concepts fundamental to the anthropological literature on Mexico, that of *community* and that of *center/ periphery,* neither of which accounts for transnational migration, though some community studies did discuss the social and spatial effects of rural-urban migration, but usually posed in opposition and as a threat to the so-called community (Cancian 1994; Redfield 1950).

Rouse's insights were further developed by others in the 1990s and 2000s who made migration, conceptualized in terms of transnational space, a defining feature of Mexicanist anthropology and of U.S.-Mexican studies more generally (Castañeda 2006; Díaz-Barriga 2008; Goldring 2000; Hirsch 2003; Hondagneu-Sótelo 1994; Kearney 1995; Lewis 2006; Stephen 2007; Zlolniski 2006). More recently, multisited ethnographic studies of the circulation of people and objects between the United States and Mexico have documented the material constitution and reproduction of transnational space, race, and inequality (De Genova 2005; Hirsch 2003; Lewis 2006; Mendoza 2006). This scholarship has emphasized the ways that a lived U.S.-Mexican transnational space is built up over time through a host of objects and practices: the movements of people; the circulation of objects such as videos, letters, souvenirs, and commodities; labor relations; and the re-creation of public and domestic space.

This emphasis on transnational space as formed through material processes involving objects and places as well as people emerges from both the commodity chain literature developed over the past twenty-five years (e.g., Chibnik 2003; Collins 2003; Mintz 1985; Myers 2002) and recent anthropological considerations of materiality and social life (Miller 1987, 2005; Myers 2002). The view of social life as constituted through people *and* things *in* places provides a strong case against the idea of "globaliza-tion" as a de-territorializing process.

W. Warner Wood's 2008 consideration of the movement of Zapotec textiles as a global process provides an example that resonates strongly with the case of Mexican minerals. Wood begins by describing a conven-tional "tour of Zapotec textiles" sponsored by the Los Angeles County Museum of Natural History, which he is asked to lead and which is launched by a pretrip slide show at the Guelaguetza restaurant in Los Angeles's Koreatown. To this he counterposes an alternative tour, one that does not exclusively focus on Teotitlán del Valle, Oaxaca, the village where many textiles are made and that is often taken as their emblem-atic and authentic source, but also includes the Guelaguetza restaurant itself, where the cultural patrimony of Oaxaca is displayed on the walls and in the music and cuisine as well as many other places. Wood points out that the restaurant should not be seen as the pretrip frame for the tour, as it was assumed to be by the tour's planners and participants, but the first stop. Nor can the restaurant be taken as a site of the "global," whereas Teotitlán del Valle stands as the "local." Rather, the restaurant and many other places participate in the ongoing production of what Doreen Massey has called "a global sense of the local, a global sense of place" (1991:29) through the movement of "textile, weaver, tourist, design, yarn and so on" (Wood 2008:25).

Mexican minerals are by definition "Made in Mexico" (to borrow Wood's title), but they are made *as valuable* in many places, including the mining localities they come from; the border cities where they cross into the United States; the gem and mineral shows in Tucson, Denver, and Munich; and museums such as the Smithsonian Institution's Natural His-tory Museum and the Harvard Museum of Natural History. In turn, they help to create those places as part of a transnational U.S.-Mexican space.

Given that minerals are objects produced by nonhuman forces that have frequently been the subjects of scientific examination, this study also contributes to an emergent field of postcolonial technoscience, which, in the words of Warwick Anderson, seeks to "understand the ways in which technoscience is implicated in the postcolonial provincializing of 'univer-sal' reason, the description of 'alternative modernities,' and the recogni-tion of hybridities, borderlands and inbetween conditions" (2002:643). Two studies have begun the project of looking at Mexico through the lens of postcolonial technoscience: Corinne Hayden's *When Nature Goes*

Public (2003) and Gabriela Soto Laveaga's *Jungle Laboratories* (2009). Hayden tracks bioprospecting in Mexico, focusing on idioms of intellectual property in postcolonial engagements between scientists, indigenous people, and the state. Soto Laveaga traces the intersection of peasants, cooperatives, scientists, and the Mexican state in revaluing *barbasco,* a wild yam rich in diosgenin, a hormone used in making the contraceptive pill. Both books look at the formation of transnational markets, the crisis in the postrevolutionary Mexican nation-state, and rural disenfranchisement from the perspective of scientific knowledge and intellectual property rights. Mineralogy and mineral collecting are intimately linked to mining and territorial expansion and contraction, which have been fundamental to the formation of the Mexican nation-state and to U.S.-Mexican relations. Studying these things thus brings up new questions in U.S.-Mexican technoscience that have not been covered in other works.

Indeed, most ethnographies of technoscience focus on organic forms and life sciences (e.g., Franklin 2007; Helmreich 2009; Lowe 2004; Matsutake Worlds Research Group 2009; Mol 2002; Paxson 2008). This study extends this approach to inorganic matter and the earth sciences. It doing so, it joins a few preliminary anthropological sorties into the earth sciences, including Claude Lévi-Strauss's (1992 [1955]) extolling of geology's attention to fundamental structures and the coincidence of forms from vastly different epochs, Kim Fortun's (2009) call for using the scalar methodology of the geological sciences as a model for ethnography and, more concretely, Andrew Walsh's (2010, 2012) work on sapphires from Madagascar.

The ethnographic emphasis on life sciences may in part be due to the fact that rocks, minerals, volcanoes, and meteorites might seem self-evident and culture-free in a way that humans, animals, and even genes and microbes do not. Nevertheless, the earth sciences engage powerful questions about the origins of life and the planet, the relationship between humans, nature, and the divine, and the material experiences of place and belonging. These are some of the reasons that people value them. In the nineteenth century, the establishment of "geological time" through mineral analysis helped to destabilize biblical accounts and the idea of God as the prime mover. People in Guanajuato and other Mexican mining localities place minerals on altars as gifts to the saints. Collectors focus on gathering minerals from corners of the globe or building collections to represent particular regions or localities. My neighbors in Guanajuato gave me minerals to remind me of—and link me to—the place once I left.

In the case of Mexico, the earth sciences play an especially important role. Mining is the main reason for this, engaging as it does highly vexed questions of state sovereignty and global markets. In more practical terms, mining brought many Mexicans and Anglos together from many different classes, perhaps more than any other pursuit (at least until the surge in labor migration after World War II). From the early years of the republics, scientists in both shared insights and research and traded minerals. Mexico was an important site for "strategic minerals" during World War II, and a good deal of scientific research was promoted in the service of the war effort (Paz 1997). The influence of the science and the social relations goes both ways: These encounters helped to shape the earth sciences, and work in the earth sciences helped to shape these encounters. All these factors make an ethnographic study of Mexican minerals particularly revealing.

The Makings of Value

One powerful motivating force for the production of material-social spaces (transnational or otherwise) is, I argue, the attempt to create value. To see how this works, we need a workable theory of value-making. This is far from a new concern within social theory. For centuries, people have tried to develop a way to analyze value cross-culturally and, as part of that endeavor, to identify what, ultimately, makes value. These discussions emerge from the history of political economy in the eighteenth and nineteenth centuries—a history that has bequeathed to us, among other things, a concern for the origins of things at the basis of social inquiry. The key question has been "what is the source of value?" The ultimate source of value has usually been identified as either labor (following Locke, Ricardo, Adam Smith, and Marx, as in much of Marxian political economy) or desire, with exchange as its enabling mechanism (following Carl Menger and Georg Simmel, as in much contemporary economics). Commodity studies, such as Sidney Mintz's *Sweetness and Power* (1985), build on the assumption that labor forms the primary wellspring of value, whereas the literature on "the social life of things" (Appadurai 1986) derives its energy from the idea that exchange is a fundamental dimension of value. These different premises lead to different research: Mintz and others, including sociologists studying commodity chains (Gereffi and Korzenewiecz 1994), tend to look at the passage of a commodity through the whole process of production, distribution, and exchange, whereas anthropologists influenced by Appadurai and his followers often follow objects through multiple regimes of value, including but not limited to commodity exchange, and in doing so seek to show how different forms of culturally constituted desire are mediated through exchange (e.g., Myers 2002).[2]

By the 2000s, there seemed to be a stalemate between those who argued for labor as the ultimate source of value and those who argued for desire or exchange, often with the implication that there could only be one true source, against which all others must be measured. To overcome this stalemate, two scholars—David Graeber and Daniel Miller—have recently advanced new ways of thinking about value and where it comes from.

In an ambitious and provocative collection of essays published in 2001, David Graeber seeks to create a theory of value that overcomes the pitfalls of earlier anthropological considerations. He argues that these earlier formulations fall either into economism, where value becomes simply the measure of desire (i.e., how much one wants something, as measured by how much one will give up to get it), or into a static, Saussurean sense of value as "meaningful difference" within a culturally defined system, without dealing with the problem of why some differences are preferred over others.

In contrast, Graeber draws on an expanded version of Marx's labor theory of value and proposes that value be defined as rooted in action. Particular cultural systems of meaningful difference can then be compared, not in terms of their internally defined characteristics, but in terms of the human action that is expended to achieve greater value within any system. Things may not be comparable, but creative action (broadly defined) is. Graeber proposes that "value emerges in action, it is the process by which a person's invisible 'potency'—their capacity to act—is translated into concrete, perceptible forms" (2001:45). He develops this idea first from Marx and then from the work of Nancy Munn (1986) on value in Gawa, an island in Melanesia close to the Trobriands, and of Terence Turner (1995) among the Kayapó in Brazil.

In the abstract to an article published in 2008, "The Uses of Value," Daniel Miller begins boldly: "This paper proposes a new theory of value based on the observation of how people use this term" (1122). In the article, he critiques what he calls "bottom-line theories of value" in which he includes both the groups mentioned earlier, at least in their more simplistic forms. What these approaches have in common is their search for a fundamental principle of commensurability (labor, desire), but Miller argues that in seeking for this bottom line they pull away from many people's everyday understanding and use of the concept to bring together different forms of value without reducing them to a common source. When people say that a piece of clothing is a "good value," for instance, they often mean it combines price, quality, and style in a successful whole, not that it is the very cheapest, best-made, or most stylish. Value, argues Miller, becomes the idiom through which people bridge the "incommensurable polarity between value as price, and value as priceless" (1122; see also Zelizer 1994). Miller draws his inspiration for this argument in part from Marilyn Strathern's (1990) resistance to the application of a labor theory of value in Mount Hagen, PNG.

In developing my own discussion of value, I borrow from both Graeber and Miller. Both have in common, in contrast with some earlier theories of value, a view of value as a dynamic process that may be concretized in different material and immaterial objects but cannot be reduced to the static *worth* of those objects (Eiss and Pedersen 2002; see also Gregory 1997). These two writers are primarily concerned with what people actually do when they make value. Following Graeber, my analysis depends on the idea that we should look at "what seem to us to be fixed objects as patterns of motion, and what seem to be fixed 'social structures' as patternings of action" (2001:xii). At the same time, I take Miller's point that a search for a unitary source of value undermines the vitality of the concept, which depends on the coexistence of irreducibly different evaluative scales.

Making Meaningful Difference and Making Difference Meaningful

In this study, I define value as the politics of making and ranking differences and deciding what kinds of differences are important. It is the negotiated process of discriminating and ranking occurring at (at least) two different levels of experience and activity.

There is (1) the process of discriminating between particular objects, ideas, or actions and ranking them within a given system. At this level, making value means comparing two *things* (e.g., two paintings), deciding in what ways they are different, and ranking those differences. Activities at this level depend on some kind of consensus about a system of qualities in terms of which things might be compared—some basis of similarity against which to discriminate and rank. We could call this *making meaningful difference*.

Then, on another level, there is (2) the process of deciding what kinds of *qualities* are worth discriminating and ranking, what differences are meaningful, or (to put it another way) of deciding what the appropriate grounds are for similarity within which objects might be contrasted. For instance, in the United States, height is a less meaningful quality for judging female attractiveness than weight, and shape of earlobes is hardly at all meaningful (at least as far as I know). Likewise, in the United States, skin color can often provoke discrimination, whereas handedness usually does not these days, though it used to. Ethnography is especially good at getting at the details of this level, which consists of identifying those things and qualities that *ought to be* important or desirable. We could call this level *making difference meaningful*. Therefore, it is not only the relative rank of objects within a system that is up for debate, but also the system of qualities itself, and *its* relative rank with respect to other systems.

The attention to the active and processual nature of value that I underscore here resonates with C. A. Gregory's concept of "valuation," which he defines as a process of comparing objects within a "generally accepted standard of value" (1997:13). Gregory goes on to note that standards of value "are generally accepted but never universally so," introducing a necessary component of diversity of opinion and contention. My account builds on this concept of valuation through close attention to the process by which what Gregory calls "standards of value" are themselves constituted as valuable.[3]

Graeber spends a good deal of energy critiquing the concept of value as meaningful difference on the grounds that "it is one thing to say that women in a market in Papua New Guinea are likely to see two lumps of apparently identical fish as different. It's quite another to say, why, as a result, a given woman will want one and not the other" (2001:43). However, I find the idea of meaningful difference to be very useful. Introducing the second level of "making difference meaningful" helps to account for why a person might prefer one object over another, thus avoiding the problem Graeber points out. This is because it draws attention to the practices

19

through which ideas of what kinds of difference should be meaningful get established. Although people often use the word *value* as a noun, saying that a given thing—say, a bottle of Bordeaux—"has value," this is actually a kind of shorthand for what really is going on: a particular bottle of Bordeaux has emerged at the intersection of many claims about how wines can be compared and which differences are better or worse (French wine is the best; people who drink French wine have good taste and can afford more expensive wine; 2005 is a good year for Bordeaux; and so on).

My proposed approach to value also shows how the systems within which objects are evaluated are also the result of clustered actions. Someone might prefer a South African wine to a French wine, in part to show that he or she has an especially discriminating palate and can see beyond the obvious prestige and price of French wine ("I'm not one of those chumps who only buys French wine"). If this happens often enough, South African origin will become a valued quality for wine. Once we approach value as a constellation of action that produces things and qualities, not as a static thing or quality in itself, we can see how all sorts of motivations, allegiances, associations, and interests intervene to create value at the two levels described previously.

Value's dynamic, generative force derives precisely from the ability to create difference. Furthermore, this feature of value is underexamined, perhaps because it seems so self-evident. For the many overlapping and crosscutting differences made between minerals and between the ways that minerals should be distinguished, I use the term *value-making acts*. These value-making acts do not end with themselves, but go on to make many other things—including collections, technoscientific objects and knowledge, spatial imaginaries, transnational relations, and markets. Furthermore, value-making acts also make people as themselves differently positioned along commodity chains and define them as different in terms of such social categories as race, class, gender, and nationality.

Knowledge, Black Boxes, and Qualification: Stabilizing Difference

The field of science and technology studies has placed a great deal of emphasis on exploring how new forms of knowledge and new entities are created and come to act in the world. The school of thought generally known as actor-network theory has been particularly interested in this topic. Emerging in the 1980s in dialogue with sociology of scientific knowledge scholars who focused on the social embeddedness of science, actor-network theory proposes a number of premises and concepts that have become more generally influential. These include recognition of the agency of nonhuman actors, commitment to a "flat" account of associations and networks, and resistance to a priori distinctions between realms of science, technology, nature, and society. Actor-network theory focuses on the ways in which heterogeneous and multiple actors (sometimes described as *actants*) come together to form associations known as actor-networks (Latour 2005).

To describe actor-networks that have stabilized to the point that their webs of association need no longer be explained or even understood, so that they act as taken-for-granted objects in the world, Bruno Latour borrows a term from engineering and computer science: *black box.* In its original use, a black box refers to an entity or program that can be used without knowledge of its inner workings, so that only its input and output are observable. When an actor-network becomes a black box, it is no longer necessary to look at its individual components, and, indeed, it often becomes difficult to recognize that it is not (or not only) a unitary entity, but a network held in place by the overlapping and crosscutting forces of multiple actors. In this sense, the concept of a black box has something in common with concepts such as *hegemony* and *doxa.* However, where these concepts imply a theory of power and ultimately explain the taken-for-grantedness of a given object or situation in sociological terms, black boxes need not be interpreted solely as manifestations of the social. They emerge through the collective stabilization of all kinds of actors and forces, human and nonhuman.

For instance, Bruno Latour begins his book *Science in Action: How to Follow Scientists and Engineers through Society* (1988) with a series of descriptive flashbacks to show us the race to model deoxyribonucleic acid (DNA) in 1951, which was ultimately stabilized as the double helix, and the building of a computer in 1980, ultimately called Eclipse MV/8000, which eventually won out over its competitors in producing three-dimensional representations of the double helix. His flashback strategy shows us moments when these bits of knowledge—the double helix, the Eclipse MV/8000—were not yet the accepted forms of describing and displaying DNA. By 1985, in his third "scene" (presented first), he reports a conversation between two scientists who describe the image of the double helix on the Eclipse computer as a "nice picture" on a "good machine."

Alternatives to these scientific facts and tools were no longer in play; the double helix and the computer had become black boxes. Latour explains:

> That is, no matter how controversial their history, how complex their inner workings, how large the commercial or academic networks that hold them in place, only their input and output count. When you switch on the Eclipse, it runs the programs you load; when you compare nucleic acid sequences you start from the double helix shape. (1988:3)

Black boxes are bits of knowledge that are "unproblematic and certain" (1988:3); that is, their problems and uncertainties have been silenced through the stabilization of a network of actors.

Although the concept of black boxes is still mostly used to describe technoscience, it has relevance beyond this field as a good way of describing how value gets produced. The stability and seemingly intrinsic character of value is the effect, not the cause, of successful claims by multiple actors. These successful claims have made the double helix meaningfully different from other shapes as the one shape that "accurately" represents the structure of DNA.

Let me give an example. In an economic anthropology class several years ago, we were discussing value's subjective nature, and one student told a story of when he was a boy and wanted his mother to buy him a particular Magic: The Gathering trading card, which cost five dollars. His mother had said, with some indignation, "I'm not paying five dollars for that. It's just a piece of paper!" Another student pointed out that the bill was also "just a piece of paper."

One could describe these two bits of paper as less and more successful black boxes. The trading card has some success (more than, say, a piece of Kleenex), for some people will see it as having meaningful boundaries separating it from the rest of the world and making it worth, potentially, US$5. They need not inquire into the circumstances by which Magic: The Gathering cards become valuable (including the efforts of the company's game-makers and designers, the gaming stores that distribute the cards, the predilections of young American boys in the 1990s, the particular qualities of dye and paper, and so on) because they have already accepted that premise. However, they probably also realize that not everyone agrees with them.

The five-dollar bill, on the other hand, is a very successful black box. Many people understand the process by which a given piece of paper is designated as legal tender (or at least know that some such process exists), but few think about it very often and it is not necessary to understand it in order to appreciate the paper's value. Indeed, the effect of its success as an object of value that is meaningfully different from a Kleenex and a Magic: The Gathering card rests on *not* thinking of the struggles that went into stabilizing those differences. The fact that neither my student nor his mother saw any absurdity in her statement "I'm not paying five dollars for that. It's just a piece of paper!" demonstrates the great stability of the US$5 bill as valuable object. However, sometimes a black box disintegrates so that the multiple actors that form it become unduly visible, calling its value into question. The skyrocketing inflation of the Zimbabwean currency in the mid-2000s is an example.

Cognate to these discussions is the concept of *qualification* as defined by Michel Callon, Cécile Méadel, and Vololona Rabeharisoa in an influential article published in the journal *Economy and Society* (2002). Callon et al. are concerned with the process by which goods are situated within markets. For this to happen, the *singularity* of a good must be established by "positioning it in a space of goods, in a system of differences and similarities, of distinct yet connected categories" (2002:198), often through highly specialized tests and measures. This positioning, an ongoing and continually readjusting process, they call qualification.[4] Through qualification, attachments between goods and consumers become stabilized and destabilized. This notion of the ongoing ratification of qualities with the object of stabilizing their position as *singular* (i.e., different from other goods in a legible, meaningful way) bears much in common with my argument and provides inspiration for further conversations between the anthropology of value and science and technology studies (e.g., as in Foster 2007).

The concepts of actor-networks, black boxes, and qualification give us ways to think about the specific mechanisms and forces through which some kinds of difference come to seem durably meaningful, to the point where they act as structures within which things (objects, people, places, qualities) come to have value.

These concepts also disrupt, in a productive way, the literature on transnational commodity chains. Much of the research and motivation for this book has consisted of following minerals mined in Mexico as they travel through Mexico and the United States, following the tradition of commodity chain studies, or accounts of political economic relations told through the movement of commodities (Gereffi and Korzeniewicz 1994; Mintz 1985). These approaches have been extremely robust at producing accounts of power relations within capitalism and colonialism, and at looking at cultural difference as manifested in different systems of value.

Even though one finds in these studies detailed descriptions of the sensible world, they are mostly deployed as devices for revealing *social* relations, especially relations of inequality. In this way, commodity studies in this political economy tradition (including my own: Ferry 2005a) base their analyses on a strict distinction between the social and natural worlds and certainly between human and nonhuman actors. They tend to approach nonhuman objects primarily as expressions of and conduits for social action and meaning. Though often committed to the idea of reciprocal causation between the material and social worlds, the main protagonists in these stories are always humans. Combined with a narrative, even biographical structure (Kopytoff 1986), the emphasis on human agents in these accounts does a good job of describing the already stabilized political economic landscape, but it is less successful at showing how that landscape came to be and how it might have been otherwise. The concepts of actor-network, black box, and qualification do not rest on this firm distinction between humans and nonhumans and, therefore, help to emphasize the ongoing, contingent character of value-making.[5]

So far I have used as examples tangible, nonhuman objects—bottles of wine, trading cards, and so on—and indeed, much of the value-making I describe in this book consists of differentiating and ranking the minerals as nonhuman objects and evaluating their presumed qualities (e.g., rarity, luster, exemplarity). However, the book also addresses the differentiation and ranking of people and places, and *their* presumed qualities. It looks at how value-making also creates people as miners, dealers, collectors, and scientists *and also as* men and women, Mexicans and "Anglo" Americans, residents of northern Mexico and the southwestern United States, white and brown, rich and poor, citizen and "alien." These different kinds of people also are presumed to embody certain qualities (e.g., masculinity, connoisseurship, laziness). Value-making creates objects as different and creates the people with whom they come into contact as different.

It is important to note that the quality of rarity ascribed to a mineral is not the same as the quality of laziness ascribed to a person or group. There are serious deleterious effects to saying "Mexicans put things off until tomorrow" that do not pertain to saying "purple adamite is rare." These kinds of difference are not the same in every way. But the process by which they are created is similar. Attending to this can show us how differences *come to seem* natural and unchanging, not only between things, but also between people and places. Through the course of the book, we will see how diverse differences have been created and dismantled over time: different mineral species, collections, places, markets, and people differently positioned along the commodity chain, and with respect to social categories of difference such as race, gender, class, and nation. Acknowledging that these are not all the same kinds of things, I suggest that we look at them as analogous in the ways they convene and disintegrate. Looking at value-making as a generative force for difference shows us just how contingent, unstable, and very consequential difference can be.

Through the rest of this book, we will follow closely how people, things, and places are created and dismantled as valuable objects. These explorations of value-making with minerals will yield an account of the United States and Mexico and particularly of the transnational spatial experiences and practices that link and divide the two countries, an account that shows us dimensions usually ignored or unseen. More broadly, this account uses the concept of value-making as the stabilization of meaningful difference to understand "the production of space" not as an abstraction but as a specific and concrete process.

Organization of the Book

My discussion centers on particular places where value-making with minerals has been especially densely clustered. In chapter 1, I describe my four most important field sites. Three of these are towns (Guanajuato and Mapimí in Mexico and Tucson in the United States) where much of my research has taken the form of interviewing and participant observation. The fourth "site" is an institution (the National Museum of Natural History of the Smithsonian Institution), and most of my research into it has been archival. Given the role of mineral museums—and of the Smithsonian in particular—in making Mexican minerals valuable, I have chosen to treat it as a field site analogous to the other three. This chapter aims to give a sense of some specific local experiences in Mexico and the United States in which the valuing of minerals is grounded.

Chapter 2 looks at two major shifts in the valuation of minerals. These shifts demonstrate both the instability of value-making and its embeddedness within larger social processes. First, I trace how mineral sciences in Mexico and the United States changed places, in the wake of the Mexican-American War, U.S. westward expansion, the building of the railroads, and the entrance of U.S. mining companies into Mexico. Second, I follow the emergence of a new dimension of value in U.S. mineral collecting since the 1970s—that of "aesthetic collecting"—with less attention to locality and scientific information and an emphasis on perfection and a "pristine" aesthetic. The aesthetic minerals movement explicitly compares minerals to fine art and has succeeded in raising mineral prices tremendously. The chapter draws on examples such as the career of the Spanish/Mexican mineralogist Andrés Manuel Del Río, who taught the first mineralogy course in the Americas in 1795, and the amazingly successful efforts of Smithsonian curator Paul Desautels to make minerals into aesthetic objects desired by a new class of collector-connoisseurs.

Chapter 3 examines the making of scientific knowledge. Borrowing from both political economy and actor-network theory, I show how particular objects of scientific value were created (or failed to be created) and how these became or failed to become new protagonists in the ongoing process of making value. I draw on three examples: the 1891 discovery of the mineral *aguilarite,* one of the few minerals named for a Mexican (rather than U.S. or European) scientist; the mineralogical activities of the two Boundary Commissions jointly responsible for surveying the new border after Mexico's defeat in the Mexican-American War (1850s); and current efforts to research and publicize the spectacular gypsum caves in Naica, Chihuahua.

Chapter 4 looks at mineral collections as a particular site of value-making. I look at how collections work as expressions of transnational space in miniature. This too is a lively process of claiming and establishing value. The subjects of my stories include a mineral collection in Guanajuato, Mexico, whose individual pieces have little value outside the collection; the Harvard University mineralogical collections, using my own donation of an acanthite (silver sulfide) and silver specimen from Guanajuato to trace the process by which a specimen enters a collection; and the most famous Mexican collection, that of Dr. Miguel Romero, which was broken up after his death in order to realize the tremendous economic value of some of the individual pieces (including the Aztec Sun mentioned on page 6). These collections express qualities of universality, Mexicanness, patrimony, and marketability, in different combinations.

Chapter 5 asks how value-making helps to create places in articulation with larger spaces. It contrasts the view of many collectors in the United States that in collecting minerals they are gathering value from distant, peripheral places into the center (a perspective fostered by world-systems theory and consistent with other imperial visions) with the view of some miners and others in Guanajuato who see the mines and Guanajuato as a central point, sending value out into the world, which then returns in other forms. Each view sees the movement of minerals as its material instantiation. Proponents struggle to valorize their own perspective at the expense of the other; in that struggle, the view of the U.S. collectors often dominates, with implications for markets, museums, and other sites.

Chapter 6 looks at the creation of markets for minerals in Mexico and the United States. I compare two marketplaces where high-end mineral specimens are sold: Mapimí, Durango, and Tucson, Arizona. The prices for minerals in these two markets are hugely different, and dealers make their living through *arbitrage,* or the strategic movement of commodities between markets with different price structures. Conventional economic theory says that over time arbitrage should undercut the conditions of its own existence, and yet the price gap has remained open over a period of decades. I examine how boundaries are policed in these markets by miners, dealers, and collectors so that only some can buy in Mapimí, and others can sell in Tucson. What can this tell us about how value-making creates differences among markets, places, and the people associated with them?

The book concludes with two stories that show with particular clarity the ways that minerals become both protagonists and narrators in the making of value.

1 HISTORIES, MINERALOGIES, ECONOMIES

Mexican mineral specimens are surrounded by webs of transactions in which the minerals, museums and other institutions, and U.S. and Mexican miners, dealers, collectors, curators, and scientists all participate. It is impossible to draw clear boundaries between Mexican mineral collecting and mineralogy and the mineral collecting and mineralogy that goes on in other places. Likewise, the boundaries may become blurred among mineralogy and geology, meteoritics (the study of meteors), and paleontology; among collecting minerals, gems, or fossils; or between those who collect ore minerals (those associated with economic ore deposits and extracted as a by-product of ore mining) or gem minerals.[1] For me, this has made the question of what to study and where to stop especially difficult.

What can we learn from limiting the object of study, for analytic purposes, to minerals as agents for making value in and between the United States and Mexico? The transactions surrounding Mexican minerals do specific kinds of work in the world: Through the process by which multiple forms of value are created, minerals help to make other things, such as scientific knowledge, collections, places, and marketplaces. And they make, to some degree, the people and places that participate in science, transnational space, and a stratified mineral economy.

Such a dispersed and weblike object of study has made necessary a particular research strategy, one that is more and more common in anthropology: multisited ethnography. Mineral specimens move, and their value is often determined in multiple places. I have conducted ethnographic and archival research in three countries and many cities, in archives, and on the Internet. At the same time, my methods have been fairly traditional: participant observation, interviewing, and archival research. In this chapter, I describe my four most important field sites—three of these are towns, and many of the activities within them are related to mining

FIGURE 1.1. Guanajuato, from Calle Potrero. Photo by Glen Sherman. Reprinted with permission.

and mineral specimens. The fourth site is an institution, and most of my research into it has been archival. Given the role of mineral museums—and the National Museum of Natural History of the Smithsonian Institution, in particular—in making Mexican minerals valuable, I have chosen to treat it as a field site analogous to the other three.[2]

I hope that an understanding of these four sites will do two things: (1) provide the reader with information necessary to understand the chapters that follow, (2) give a sense of some of the specific local experiences in Mexico and the United States in which the valuing of minerals is grounded.

Guanajuato

The city of Guanajuato is the capital of Guanajuato state, located in the center-west region of Mexico (figure 1.1). The city is the seat of the *municipio* of Guanajuato, which reported a population of 153,364 in the census of 2005 (this includes a number of communities outside the city). Guanajuato is one of the few state capitals in Mexico that is not also the largest and most influential city in the state (nearby León had a population of 1,278,087 in 2005).[3]

The Guanajuato mining district is located in the central part of Guanajuato state. Local legend has it that silver was discovered in Guanajuato in 1548 by muleteers on their way back from Zacatecas (Marmolejo 1988 [1886]). However, the scale of exploitation was relatively small until 1768, when a great bonanza was struck at the Valenciana mine. This discovery transformed the history of the city. Since that time, the Guanajuato district has produced over one billion ounces of silver. David Brading reports that "at the close of the eighteenth century Guanajuato was the leading producer of silver in the world. Its annual output of over 5 million pesos amounted to one-sixth of all American bullion, gold and silver combined, and equaled the entire production of either viceroyalty of Buenos Aires or Peru" (1971:261). The majority of ore is found in the *Veta Madre* (Mother Lode), a vein system of silver-ore-bearing quartz and calcite that runs 12.9 kilometers from northwest to southeast, ranging from 18 inches to 160 feet wide (Panczner 1987:38–39).

In the second half of the eighteenth century, then, Guanajuato was the leading silver mining center in Latin America, a position it inherited from the mines of Potosí in colonial Peru (in what is now Bolivia) in the sixteenth century and from Zacatecas in the early seventeenth century (Bakewell 1984; Brading 1971). Silver made Guanajuato and the surrounding Bajío region into an engine for global markets and a bustling capitalist system in its own right (Palerm 1980; Tutino 2011). However, the industry was severely damaged in the War of Independence that began in 1810; the mines were abandoned and allowed to flood (Rankine 1992; Ward 1828:441). In 1825, the Valenciana mine was purchased by the Anglo-Mexican Mining Company, which drained the mine and instituted steam power. However, the costs of drainage and the absence of new methods for processing ore inhibited production, and the company shut down in 1848 (Rankine 1992:31). The district revived with the bonanza of the mines of nearby La Luz in the 1840s (Blanco et al. 2000:124; Jáuregui 1996) and again with the advent of cyanide processing in the first years of the twentieth century (Martin 1905; Meyer Cosío 1999; Rickard 1907). U.S. mining companies began to arrive in the last several years of the nineteenth century, and their presence increased dramatically after the arrival of electricity to Guanajuato in 1904. Between 1897 and 1913, about seventy mining companies operated in Guanajuato, the vast majority of them U.S.-owned (Meyer Cosío 1999:101).

In the 1930s, a series of strikes (part of a burgeoning national labor movement and an upsurge of resource-based nationalism) disrupted most of the mining corporations in Guanajuato. In particular, the Guanajuato Reduction and Mines Company, which had been one of the largest companies in the district since 1904, was forced to leave on account of low silver prices, declining yields, and a series of strikes. The company ceded the holdings to the workers in lieu of its debts. In June 1939, the workers, formerly organized as Section 4 of the National Miners' Union, reorganized as a producers' cooperative, the Sociedad Cooperativa Minero-Metalúrgica Santa Fe de Guanajuato (hereafter "the Santa Fe Cooperative").

The Santa Fe Cooperative operated from 1939 to 2005, when it sold most of its surface holdings to a Canadian company, Great Panther Resources (now Great Panther Silver), Limited, through a Mexican partner, El Rosario. During its rollercoaster life, the cooperative went from penury and crisis in the 1940s (when an engineer from Mexico City was brought in to run it), slow stasis in the 1950s–1970s, and a boom period in the 1980s, when the price of silver skyrocketed from US$8 to US$50 an ounce (Ferry 2005b; Jáuregui 1985).

Along with the cooperative, two other companies operated in Guanajuato: the locally owned and managed El Cubo Mining Company and the Grupo Guanajuato projects owned by Peñoles, one of Mexico's largest mining corporations. Peñoles came to the city in the 1960s and operated a number of mines through the 1990s, largely with contract labor. During the 1990s, when I was doing field research on the cooperative, silver prices were low (between US$4 and US$7 an ounce) and the future of mining appeared grim. Many men were leaving the city to work on contract for mines in the north of Mexico or to migrate to the United States.

Precious metals prices have skyrocketed over the past decade, particularly since 2005. The yearly averages for silver and gold in 2000 were US$4.95 and US$271.04, respectively; in 2005, they were US$7.32 and US$444.74, and in 2012, US$31.15 and US$1668.98.

These price rises mean that exploration for, and mining of, gold and silver have sharply risen the world over, and in Mexico in particular. In Mexico, mining has been dominated in recent years by Canadian companies. Although no complete list of Canadian mining projects currently operating in Mexico exists, there are certainly several hundred at various

31

stages of exploration and production. Great Panther is one of dozens of "junior" mining companies operating in Mexico. Junior companies are relatively smaller than giants such as Newmont, BHP Billiton, and Goldcorp, and junior companies depend to a greater extent on investment and venture capital.

Since 2005, Great Panther has restructured the workforce, only about 20 percent of whom are former cooperative members (according to sources within the company). The company has also brought many workers in from outside Guanajuato. It has rebuilt many parts of the central plants and recapitalized the mines. The company website reports that the Guanajuato mine complex milled 144,000 metric tons of ore and produced 1,019,856 ounces of Ag (silver) and 6,619 ounces of Au (gold) in 2010 (Great Panther Silver Ltd. n.d.).

Guanajuato is one of two sites wholly owned by Great Panther (though, in keeping with Mexican law, the rights to exploit subsoil resources are not owned outright but are leased from the federal government). In the past few years, the company has also carried out exploration in Chihuahua, Chihuahua; Mapimí, Durango; San Antonio, Chihuahua; and the Sierra de Santa Rosa, Guanajuato.

Great Panther has been criticized in Guanajuato for refusing to pay out the yearly 10 percent profit shares required by Mexican law for illegal dumping of mine tailings into a local reservoir and for poor safety conditions. In this respect, it has not received as much criticism as AuRico (formerly Gammon Gold), which operated the El Cubo mine until 2012. El Cubo was on strike for nine months in 2009–2010 to protest ten-hour workdays and the failure to pay profit shares (the company claimed a lack of profit at the El Cubo mine).

Minerals of Guanajuato

In the words of a group of mineralogists, specialists in the Guanajuato district:

> The mines of the Guanajuato district in Mexico are well known to collectors for producing fabulous specimens of silver sulfides and sulfosalts, calcite and amethyst. The district has been in nearly continuous production since its discovery in 1548, and has had a total silver production in excess of 40,000 metric tons. The district is dominated by three major northwest-trending vein systems: the La Luz (the western side of the district), the Veta Grande (home to the Rayas, Valenciana and Cata mines)

and the Veta Sierra (the eastern extent of the district, and home to the Peregrina and Santa Catarina mines). . . . The ore is in quartz/calcite veins controlled by northwest-trending faults, and in large stockworks. . . . The veins are rich in a suite of sulfides that include pyrite, sphalerite, tetrahedrite, arsenopyrite and cinnabar in addition to the silver mineralization. (Francis et al. 1999)

In lay terms, this means that as miners drill and blast for silver and gold ore, they dislodge quartz, calcite, and other *gangue* minerals (i.e., non-economic minerals that surround ore minerals in the veins) as well as silver-bearing minerals such as acanthite, pyrargyrite, polybasite, stephanite, and others (some of which are very rare). Miners (especially drillers) can then gather up the loose minerals or work them free with a crowbar. Miners then keep them for their own uses, give them away, or sell them, usually first to others who work in or near the mine who have dedicated themselves to building commercial networks.

The cheaper minerals, especially white quartz, amethyst, calcite, and pyrite, are sold to tourists on the grounds of the mines, at the swanky Hotel Santa Cecilia, and near the Mercado Hidalgo in the city center. These are usually sold on an occasional basis and do not form part of long-term commercial relationships between buyers and sellers (though some calcites and quartzes can also sell for higher prices to more discriminating buyers). They may cost anywhere from 20 to 200 pesos (US$1.80 to US$18.00, approximately) per specimen, and they are usually intended to be used as souvenirs or decorative objects.

The silver minerals tend to bring higher prices. These are usually not sold publicly, but through personal trade networks, often with the financial backing of dealers from the United States. In contrast to the cheaper gangue minerals, sales are usually part of business arrangements that may last for months or years. Minerals have been sold in Guanajuato since at least the late nineteenth century. The mineral collector Ponciano Aguilar recorded prices of minerals from the San Carlos mine beginning in the 1880s. Because he was the mine superintendent, it seems logical to suppose that he bought these minerals from the miners. However, the popularity of minerals as tourist souvenirs dates from the development of Guanajuato as a tourist destination in the years up to and following World War II. A merchant known as Santitos set up a stand near the Plaza San Roque in the 1930s; he is reported to have been the first to sell minerals in a regular way. During the 1960s–1980s, many merchants sold minerals (and other tourist goods) near the bus station on the Avenida Juárez. When the bus station was moved to the outskirts of town in the early 1990s, this trade began to die down.

However, over the second half of the twentieth century, as in other Mexican mining localities, a small but enduring network of mineral dealers and suppliers developed. George Griffiths of Gómez Palacio traveled periodically to Guanajuato with his wife, as did some other North American dealers. In the 1990s, Christopher Tredwell, a British auto executive living in the city of Irapuato, not far from Guanajuato, obtained permission to enter the cooperative mines and self-collect, as well as to purchase specimens from miners. His collection was eventually sold to the Harvard Museum of Natural History and the University of Arizona Mineral Museum.

As a prominent silver locality, Guanajuato attracts the attention of silver collectors, who are a faithful group, on the whole. These collectors may seek either native silver (in its elemental form) or compounds that include silver (e.g., pyrargyrite, stephanite). Because they come from a "classic" locality, Guanajuato specimens continue to have an appeal. Specimens from there typically cost in the hundreds rather than the thousands of dollars depending on a host of characteristics, such as species, form, size, and so on. Just to give some exemplary prices for Guanajuato minerals, in February 2011, one could find for sale at various dealers' websites a calcite and pyrargyrite specimen (8.5×6×2.5 centimeter) for US$750, an amethyst and calcite specimen (6.8×4.7×3.1 centimeter) for US$50, a thumbnail stephanite piece (10 mm long) for US$500 and an acanthite specimen (of unspecified size) for US$1,250. (One can see here that the amethyst piece is much cheaper than the specimens with silver minerals—as would be true for stephanite and amethyst from anywhere else as well.)

Tucson, Arizona

The new visitor to Tucson notices first its topography (figure 1.2). The city is flat in the center, with rows and columns of streets with stucco houses, low to the ground, and fronted by grass or cactus gardens. The richer part of town is to the north of the city in the foothills. Driving north on one of the avenues such as Swan or Craycroft, one passes many winding roads that snake further into the hills, often with a sign at the entrance discouraging those who do not live there or know anyone who does. These roads meander like rivers to low ranch-style houses sprinkled

FIGURE 1.2. Skyline of Tucson, Arizona, after sunset, from Sentinel Peak Park.
Bigstock.com. Reprinted with permission.

through the hills, punctuated by neighborhood watch association signs.
Many have gardens with cacti and bougainvillea. There are a few scrub
trees but little real shade, unless you happen to be a lizard.

Many of the interviews I had with collectors and dealers in Tucson
took place in these neighborhoods, in the foothills of the Santa Catalina
Mountains. For one interview, I drove into a neighborhood whose streets
were all named after missions founded by the Jesuit priest Father Eusebio
Kino in the seventeenth century: Tubutama, Caborca, Busanic, Sonoyta,
Imuris, Cocóspera. Of the nine missions that Kino founded, two are in
Arizona (San Xavier del Bac and Tucumarori) and the others are all in
the state of Sonora, Mexico. The mapping of both Arizonan and Sonoran
place-names to an Arizona neighborhood exemplifies one aspect of the
twinned, but also intensely unequal, relationship between Arizona and
Sonora.

The current state of Arizona occupies the northern part of the territory known in the colonial period as la Pimería Alta (Upper Pima region),
which was taken over by the Spanish in the middle of the sixteenth century. Father Kino, who established missions in the Gila River Valley in
the late seventeenth century, noted the presence of mineral resources in

the area (Raymer 1935:123), and there were small-scale mines in the early nineteenth century, but mining exploration only began in earnest in the middle of the nineteenth century, during the California Gold Rush and just after the United States gained a vast swath of land in the Treaty of Guadalupe Hidalgo in 1848.

The area south of the Gila River, where most of the copper and silver mines in Arizona were eventually found, was acquired by the United States through the Gadsden Purchase (known as the Mesilla sale in Mexico) in 1853. This purchase was motivated in part by inconsistencies in the Disturnell Map used in drawing up the Treaty of Guadalupe Hidalgo in 1848 (Rebert 2004), which led to great confusion when the binational Boundary Commission tried to survey the new border. In addition, the United States greatly desired the acquisition of territory south of the Gila River in order to provide land for a southern railroad route across the continent.

Mining in Arizona began with gold placer mining (the mining of gold from stream sediments) and shifted to underground silver mining in the 1870s, with Tombstone as the most profitable mining center. Beginning in the 1850s, copper was being mined at Ajo, but the copper boom did not hit until the 1880s, when most of the main copper camps were formed, in Clifton-Morenci, Globe-Miami, Bisbee, and Jerome. All of these mines but Jerome are part of an extended zone of porphyry copper deposits in southern Arizona and northern Sonora (Bideaux and Wallace 1997). Porphyry copper deposits, generated by the flow of hydrothermal fluid containing small quantities of metals through hairline fractures in a porphyritic zone, are typically large deposits of disseminated, low-grade copper ore (Evans 1997:137–138). They are generally exploited through bulk mining techniques (mining of large quantities of low-grade ore). Half of the world's copper is found in such deposits, and the Arizona-Sonora region has the world's densest concentration of them (Misra 2000:372).

In 1877, silver was struck in Tombstone, but copper quickly became the major ore exploited in the area, in the mining centers of Bisbee-Douglas, Globe-Miami, Clifton-Morenci, and several others. In 1877, the Copper Queen mine in Bisbee was founded, and for decades thereafter it was Arizona's richest copper mine. In 1880, the transcontinental Southern Pacific route reached Tucson, allowing copper to be shipped east. In 1893, Nicholas Tesla invented the alternating current electrical system, creating a greatly increased demand for copper wiring. It was at this point that mining for copper became a truly profitable business (Arizona Mining Association 2006:2). Tucson, then, became the center of a regional economy dominated by copper. By the 1890s, Phelps-Dodge became one of the dominant companies in the state, a position it maintained until the 1980s when copper prices fell worldwide.

From the beginning of the industry, Mexican workers, either those migrating from the south or resident from before the 1853 Gadsden Purchase, or Sale of La Mesilla, formed a large part of the workforce. In the Arizona mining camps, labor quickly became divided along racial or national lines. Some camps, designated as American or white man's camps, had few or no Mexicans working underground, whereas others, called Mexican camps, did allow Mexicans in the better-paid and more prestigious jobs. Apparently, almost the only times in which Mexican and white workers came together were when both groups sought to exclude Chinese workers from mining jobs (Huginnie 1991:43–45). Furthermore, in most Arizona mining districts, a dual-wage system developed, where Mexicans were paid one-half to two-thirds of what whites were paid for the same labor. This system prevailed throughout the region until the mid-twentieth century (Huginnie 1991:57).

The copper mines of Bisbee, Globe, Morenci, and elsewhere produced phenomenal specimens of the copper minerals, including malachite, azurite, cuprite, and native copper, and these became part of the mineral collections of a number of Arizona engineers and others associated with the mining industry or with mining towns.[4] For instance, the Smithsonian holds an "outstanding collection of Copper Queen minerals obtained from the late Dr. James Douglass (mining engineer and president of Phelps-Dodge) as well as the T. Sterry Hunt collection of minerals" (*Curators' Annual Report* [CAR] 1944–1945). More recently, the Los Angeles County Museum received a marvelous collection from the children of Benjamin Frankenburg, the owner of a dry goods store in Bisbee from 1898 to 1928, who purchased mineral specimens from Bisbee miners (Kampf 2006). Arizona minerals, particularly from Bisbee, are highly valued these days, especially because high-quality specimens are no longer found in the mines (Bideaux and Wallace 1997).

Copper continued to be a significant part of Arizona's economy in the twentieth century. In the 1990s, the Morenci mine became the biggest open-pit mine in the country and one of the biggest in the world. In 2007, Phelps-Dodge was acquired by the corporation Freeport-McMoran

Copper and Gold, which now owns 85 percent of the Morenci mine (Sumitomo Corporation owns the remaining 15 percent).

At the same time, the economy of Tucson in the twentieth century was not entirely organized around copper. Like other southwestern cities, Tucson has grown tremendously in the past sixty years, from 120,000 in 1950 to 1,000,000 in 2006. The largest sectors of the economy are tourism and services, education (including the University of Arizona), and light manufacturing (dominated by Raytheon Corporation). Tucson is far smaller than other Western cities such as Dallas, Los Angeles, or Phoenix and maintains a small-town (or at least small-city) charm that is less apparent in these other places. Its maintenance of over 100,000 acres of desert preserve and its reliance on groundwater and a gradual introduction of Colorado River water into the municipal water supply have helped to mitigate some effects of urban sprawl (Logan 2006:191–194).

Tucson: "The Main Show"

The Tucson Gem and Mineral Show is the culminating event of a cluster of over 40 gem, mineral, and fossil shows that take place in the first half of February each year. It is a nonprofit show, put on by the Tucson Gem and Mineral Society (TGMS). Collectively, the shows are now officially known as the Tucson Gem, Mineral, and Fossil Showcase. The showcase, climaxing in the Tucson Gem and Mineral Show in mid-February, has gained nearly mythic status among mineral collectors. In 2004, a special supplement to the *Mineralogical Record* was published, titled *The Tucson Show: A Fifty-Year History,* that narrated its meteoric rise from humble beginnings. The book, written by longtime TGMS member Bob Jones, begins with an introduction by Wendell Wilson, stating: "No other event in history has tied the gem and mineral world together like the Tucson Gem and Mineral Show" (Jones 2004:7). This is an extraordinary statement, considering that mineral and gem collecting have been going on in some form for much of human history and in a more recognizably modern way since at least the sixteenth century (Wilson 1994). However, it is not an exaggeration. The Tucson shows have concentrated mineral and gem collection into one place and one time of year in a way that no other event or institution had done before. Anthropologist Hecky Villanueva writes,

> The Tucson Show is simultaneously a show, museum tour, exhibition, market, bazaar, swap meet, convention, conference, workshop, party, fiesta, pow-wow, food-fest, and tourist destination that brings together

over 50,000 unique visitors, curators, collectors, dealers, buyers, scholars, enthusiasts, tourists, students, artists, even hippies to this three-week event. (2007:1)

This does not mean that other events in other places, such as the Munich show in October or the Denver show in September, do not attract great activity, nor that mineral collecting and dealing do not go on in thousands of other places—including the virtual spaces of the Internet—throughout the year. I have tried to describe those activities in chapter 2. But the Tucson shows tower above all of these in both actual and perceived importance. They have achieved an iconic status as the pinnacle event of the year and have made Tucson a pilgrimage site for serious mineral collectors.

The shows are also extremely important contributors to the Tucson economy in terms both of direct revenue and of taxes. According to a report commissioned by the Tucson Convention and Visitors' Bureau, "the 2007 Tucson Gem Show featured 49 individual shows and 5,079 exhibitors. The total gate attendance was 363,816 Buyers, who attended 6.6 shows (on average), for an estimated attendance of 55,056 unique persons" (FMR Associates 2007:Executive Summary). The report, which primarily focused on the tax and sales revenue for the city, demonstrated a steady rise in attendance at the shows and expenditures since 2000. In 2007, it was estimated that US$9,057,217 were paid in local taxes (including sales tax and tax on car rentals and hotel rooms), up 51.2 percent from 2000 revenues (FMR Associates 2007:Executive Summary).

The original Tucson Gem and Mineral Show, sponsored by TGMS, takes place at the end of the second week in February (at the tail end of all the shows). Unlike most of the others, this show is not-for-profit; the revenues are rolled back into the show and into various community activities and projects that the society participates in (the Arizona State Museum, University of Arizona Mineral Museum, Southern Arizona Regional Science Fair, Arizona Boys' and Girls' Clubs, and the Arizona-Sonora Desert Museum) (Villanueva 2007:2).

Originally the society's show was held in an elementary school in Tucson. For the first few years, it was a small affair, made up of people from the university and collectors, especially self-collectors (for at that time the opportunities for self-collecting in Arizona mines were much greater). In the early 1960s, the society invited the Smithsonian to participate, not knowing that as a matter of policy the Smithsonian minerals did not travel. However, the curator at the time, Paul Desautels, came anyway, with an exhibit. He was welcomed warmly into the homes of the society members, and he apparently had a wonderful time. His presence encouraged other museums to come and brought legitimacy to the show. It also furthered an interest in the educational aspect of mineral collecting that was important to many of the members, part of a postwar Arizona generation without a high degree of formal education, but with a great deal of respect for learning and science.

In a letter to his friend and Smithsonian benefactor Marion Stuart, in March 1973, Desautels remarked on the growth of the show and its effect on the mineral collecting and museum world:

> Dear Marion:
>
> Our paths at Tucson certainly didn't cross much. The larger and better that show gets the more demands there seem to be on my time. On my bad days I often think of the early times in Tucson when it was possible for me to see the show and to stroll around chatting with my friends. On my good days I'm ecstatic that the Tucson Show has done so well, that the state of mineral museums in the U.S. is healthier than ever, and that my job is so exciting. C'est la Vie! (Desautels 1973)

The show eventually outgrew its school location and moved to the Pima County Fairgrounds, then to the old convention center, and finally to the new convention center in the center of the city. Over time, satellite shows in various hotels and other venues sprang up, each attempting both to capitalize on and to preempt the TGMS's show by starting slightly earlier. The event then grew a day at a time, always moving backward from the middle of February, until now the earliest shows start in the last week of January. The satellite shows occupy hotels all around the city and focus on a vast range of countries, types of collectibles (e.g., fossils, gemstones), or categories of buyers (e.g., high-end, low-medium-end, wholesale tailgaters, etc. [Ross et al. 2010]).

The rise of the Tucson show and the growing market for minerals from Mexico are closely connected. One mineral dealer with deep ties to Mexico stated that

> Mexican minerals are clearly responsible for the Tucson show becoming the Tucson show. There are mineral shows all over the world. But every other day there was a new find coming out in Mexico coming to Tucson and El Paso. And they would go to the Tucson show. You could go to

other mineral shows and see some nice minerals. But Tucson had brand new batches of phenomenal minerals in quantity. In 1962–3 it seemed like there was a never-ending supply.

Another said, in response to a question about the role of Mexican specimens in the rise of the Tucson shows:

All of a sudden, in the late 60s, there were these spectacular lead arsenates, colorful legrandites, wulfenites, and things like that. They were spectacular compared to what you see in copper country, even in Southern California [localities abundantly represented at other mineral shows—and at Tucson]. There were hundreds of flats—they bootstrapped the whole operation. There were species that had not been found before in such exquisite form and color.

These colorful minerals began to come into the United States, especially through dealers in El Paso such as Manuel Ontiveros and Colonel Baron (known for hiring goons to beat up rivals attempting to bring minerals into the United States) to dealers in Tucson, Arizona, including Suzie Davis, Richard Bideaux, and others.[5]

A new generation of dealers and collectors came up in the 1970s. Inspired by the minerals they saw at the Tucson shows and the opportunities for self-collecting in Arizona mines, they opened up new sources of Mexican minerals and established a new generation of mineral transactions between Mexicans and gringos. Many of these people are now among the top dealers in the mineral world today.

At the same time, the role of Mexican specimens in the Tucson shows should not be overstated. The showcase brings together minerals and people from all over the world, and Mexico has no more or richer localities than many other countries, including Brazil, Namibia, and Afghanistan. When one walks through the aisles of the show and looks at the displays, Mexico is not primarily what jumps to mind. More immediately present, even overwhelming, is the sense of the world's bounty of breathtaking pieces of rock, in every possible color, size, and crystal form, representing nearly every country and mineral species, their stony flames burning on the floor of the large, somewhat drab City Convention Center. Mexican minerals, however, seem to have played a formative role in the way that the show developed, and these minerals certainly make a strong showing in the various exhibits. Every year, the TGMS awards the Miguel Romero Prize to the collector with the best Mexican specimen on display, which is one of only two awards linked to a locality or localities. The other is the Richard Bideaux Award for best Arizona specimen.

In June 2008, I interviewed Carole Lee, who is a member of the TGMS show committee, which consists of about thirty members, all volunteers (the society also has two paid staff people, who spend a good deal of their time working on the show). Carole has been in charge of publicity for many years. She spoke about the difficulty of maintaining the sense of distinctiveness for the TGMS's show over the forty-odd shows going on in the vicinity, all of which except one—the American Gem Traders Association show—are for-profit shows, and all of which postdate the rise of the TGMS's show. The TGMS's show, which takes place during the second weekend of February, is the last show to open, and Carole remarks, "that makes it very difficult for publicity. When the shows begin, the press comes to me for information. The deal is that they mention our show, but we're not open yet so they can't include any photos. . . . This year (2009) for the first time we're calling it 'the Main Show.'"

For the TGMS, this struggle to maintain the show's cachet has been ongoing since the other shows began to crop up, but it does not seem to have damaged the revenues or the reputation of the Main Show. In February 2009, after the crash of the stock market in fall 2008, I visited the Tucson show, expecting to hear a lot of complaints about the lost revenue. Instead, many people spoke with surprise about the high prices and brisk sales at the Main Show, though business was off more markedly at some satellite shows such as the InnSuites show. One informant told me he knew of at least five sales for US$1 million or more at the TGMS's show. However, when I reported to another dealer that people had been talking about how well things were moving, he smirked and said, "Anyone who's telling you he had a good show is lying." Although this may well be true, of all the shows at Tucson in 2009, the TGMS's show appeared to be the most profitable and lively. As one TGMS show committee member put it, "We're the middle of the snowball, and I think, actually, if things start to unpeel, being at the core is a good place to be. Because we'll still be there as the snowball melts."

The Tucson show, as *the* event in the mineral world," was at least partially underwritten by the flow of Mexican minerals into the Southwest in the 1960s and 1970s. At the same time, the Tucson show works as a marketplace not only for minerals, but also for the criteria by which minerals are valued, of which locality is only one (and for some a decreasingly important one). The show's importance in our story arises from four features: (1) its Mexican connections, (2) its outgrowing the history of those connections, (3) the ways those connections are erased and forgotten, and (4) the ways they are displayed and remembered.

Mapimí

From my notes, August 2007:

> I take the Autobus de la Laguna from Bermejillo, which crawls along the hot asphalt road, passing a few dirt roads leading to ejidos and ranchos, a solitary cemetery and then the turn-off to Ojuela, with a ticket booth and a rock shop on each side of the road. A bit further we pass a dusty hill of what I am later told is mostly arsenic. Then the town of Mapimí begins, abruptly, with a square with a bust of Pancho Villa, and the ruins of the old Peñoles headquarters off the road to the right. A few young, very strong-looking men sit outside the building at the entrance to town. (Later I learn they are miners, many of whom live in this neighborhood [*la colonia mecánica,* whose houses were built by Peñoles and given to miners' families when they moved down the hill from Ojuela in the 1930s].) One road leads straight from the edge of town to the center, with a classic tree-lined plaza, a number of plaques commemorating the arrival of electricity and other events, and a handful of stores (hardware, video games, Pemex station, groceries). The buildings here are clearly colonial, baked and crumbling, reinforced with concrete at key points.

The town of Mapimí, Durango, was founded in 1598. It lies at the eastern edge of the region known as the Comarca Lagunera, an area of 44,887 square kilometers that occupies part of the Mexican states of Coahuila and Durango. The Comarca Lagunera also contains the twin cities of Torreón (Coahuila) and Gómez Palacio (Durango).

The region was occupied by a number of nomadic groups before the arrival of the Spanish. It was sparsely occupied in the colonial period (with the exception of Mapimí), and the towns and haciendas in the area were subject to periodic attacks by native groups. In the Porfiriato, the region became the site of industrial and agricultural development (particularly cotton production) and was linked to Mexico City and the United States through a system of railroad lines that ran primarily north-south (Hoke 2006).

The Ojuela ore deposit, near Mapimí, was discovered by the Spanish in 1598 or 1599. Historically one of Mexico's most productive mines, Ojuela is connected underground to seven other mines (América Uno, América Dos, Guadalupe, San Juan, Monterrey, Tiro Tres, and Socavón) encompassing over 450 kilometers of underground tunnels (Panczner 1987:25). The mines produce hydrothermally deposited, polymetallic (particularly lead, silver, copper, and zinc) ore from four distinct mineralized zones, of which the oxidized silver-lead and zinc-lead zones provide most of the minerals of interest to collectors (Moore and Megaw 2003; Panczner 1987:29). The Spaniards mined until they were expelled from the country in 1827, after which local owners mined sporadically until the 1860s, when the Durango-Mapimí Mining Company, incorporated in Council Bluffs, Iowa, took over, smelting up to twenty tons a day. However, the lack of transportation infrastructure made it impossible to make a profit.

In 1887, a small mining concern called Industrias Peñoles took over the workings of the Ojuela mine, establishing offices and building a processing plant in the nearby town of Mapimí. The company also built a 17-mile railroad to meet up with the Mexican central line at Bermejillo; a suspension bridge, second only in length to the Brooklyn Bridge when it was built (and indeed it was built by the firm of John A. Roebling's Sons, which also built the Brooklyn Bridge) (Hoke 2006); workers' housing near the mine itself; and housing for managers near the smelter on the eastern edge of Mapimí. The mining camp included a school, hospital, cantinas and gambling houses, a theater, and a baseball field. This last was the site of a tragedy in the first decade of the twentieth century, when the train derailed, launching four cars into the baseball field, killing forty people, including the team (Panczner 1987:22). In the managers' neighborhood, where the Americans and Europeans lived, the company also had offices and a social club with tennis courts and a swimming pool. Figure 1.3 shows Pancho Villa on a visit to the club in 1920. The ruins still stand at the east end of town.

In 1911, the miners at Ojuela went on strike and were put down by police (Knight 1990:26), and production all but ceased during the Mexican Revolution of 1911–1920, but revived during the 1920s. It is estimated that between 1893 and 1931, 3.8 million metric tons of rock were mined at average ore grades of 3.7 grams/ton Au, 462 grams/ton Ag, and 14.9 percent Pb (Hayward and Triplett 1931); the old on-site smelter having closed down, the ore was transported by railroad to the Peñoles smelter in Torreón. In 1931, Peñoles ceased direct operations at Ojuela because of low prices for metals and rising water in the mines. At that time, the

FIGURE I.3. Pancho Villa center at Peñoles headquarters, Mapimí, Durango, 1920. Collection of George E. Hoke, Bellevue, Washington. Reprinted with permission.

company moved its headquarters to the city of Torreón, and many of the workers moved from Ojuela to Mapimí (the town of Ojuela now exists only as a field of crumbling houses) (figure 1.4).

In 1940, a producers' cooperative, the Sociedad Cooperativa de Producción "Mineros de Ojuela" was formed, with fifty-seven members. Most of them lived on the eastern edge of town and scraped out a living selling scrap metal, lead carbonate to be used in the Torreón smelter, and, increasingly, mineral specimens. The cooperative controlled formal access to the mine, and members enjoyed the right to collect mineral specimens as an extra benefit of membership in the cooperative. The former director of the cooperative told me that as a member of the cooperative "after you worked eight hours, you could have two more hours to collect *riscos* [minerals]."

Like other Mexican mining cooperatives (Ferry 2005b), this one guarded its boundaries and the rights of its members jealously. One informant told me of his struggles to be accepted as a full member of the cooperative rather than working only as an "extra" without full privileges. As mineral specimens became more valuable, the right to extract them became more hotly contested. A group of *risqueros* (high-graders, or mineral specimen prospectors) who were not part of the cooperative worked

FIGURE I.4. Town of Ojuela, late nineteenth century. Collection of George E. Hoke, Bellevue, Washington. Reprinted with permission.

around the margins of the cooperative, slipping into the mine through disused adits at night or after cooperative shifts to extract riscos. Because cooperative miners would set explosive charges at the end of their shift, returning on the next shift to extract the minerals, some risqueros would come in right after the shift to check for riscos. Indeed, the Aztec Sun, the fantastic legrandite found in 1977, was extracted by Felix Esquivel, a risquero who entered the mine fifteen minutes after the cooperative miners had left and located the pocket. (Enlarging the pocket and carefully extracting the legrandites inside took a further six hours of painstaking labor.)

However, a fifty-odd person cooperative and some risquero work could not support the working population of an entire town, and the population of Mapimí plummeted in the middle of the twentieth century. During the period 1910–1940, it fell from 8,204 to 2,204. When the cooperative closed its commercial mining operations in 1981, things became even more desperate. Many men and, indeed, entire families moved to Torreón, Ciudad Juárez, or the United States in search of work. Common destinations in the United States were Denver, Michigan, and Virginia, where migrants found work primarily in *la carpeta* (carpet-laying). In this respect, Mapimí resembles many towns and small cities in northern Mexico.

A few local and regional opportunities have emerged in recent years. In the 1980s, Miguel Villegas, the owner of the Gómez Palacio company Granjas el Trasgo, established a program with local ejidos to set up poultry layer and breeder farms, producing eggs and chicken meat. In 1997, Trasgo joined with other Mexican companies to become Tyson de Mexico, producing eggs and meat for Tyson Chicken in the United States (Villareal et al. 1998). Trasgo/Tyson became one of the major employers in the Mapimí region. In 2007, workers in Tyson farms earned approximately US$80 per week. Contracts to incubate eggs and raise chickens and the reopening of the mines in the 2000s have caused a partial recovery of population (in 2005, the town's population was 4,765).[6]

In 2004, a Monterrey businessman acquired the contract to mine at Ojuela. The mine is largely extracting galena (lead sulfide) for sale to the Peñoles foundry at Torreón, where it is used to help process the vast amounts of silver the company is extracting from Fresnillo and other mines. The most experienced miners are former members of the ore-mining cooperative, some of whom also participated in the specimen cooperative in the 1980s and 1990s. Many of the younger miners are sons of cooperative members or of workers from the era when Peñoles mined Ojuela directly.

Conditions in the mines are far more rudimentary than they were in the period from 1896 to 1931. There is no electricity, and indeed no mechanized equipment of any kind; in some parts of the mine, mules help pull the mining cars. Miners work in teams; the *tumbador* extracts the galena with hammer and chisel, while *cargadores* carry it out in sacks and on flatbed cars (figure 1.5). They are paid by the kilogram, and many carry out five hundred to eight hundred kilos per day. The pay is good, especially compared with pay at the only other significant business in the region, the chicken farms contracted by Tyson.[7] And miners supplement their wages by selling specimens, as they have done for at least the past fifty years. The town has never regained its former size, but the arrival of chicken farming in the 1980s and the reopening of the mines in the 2000s have caused a partial recovery of population (in 2005, the town's population was 4,765).

Mapimí: The Making of a Market

Specimen mining at Ojuela has been a profitable concern because of the mine's extraordinary mineralogy. The mine contains copper, lead-

FIGURE 1.5. Miners pushing car, holding carbide lamps. Photo by Elizabeth Ferry.

zinc, and lead-silver, each of which has significant secondary mineralization. These secondary assemblages are where one finds highly colorful and crystallized specimens. The 120 or so mineral species from the deposit include adamite, legrandite, paradamite, scorodite, hemimorphite, plattnerite, aurichalcite, rosasite, fluorite, calcite, and wulfenite, all of which are found as superb specimens (Moore and Megaw 2003).

In the late 1940s, a handful of U.S. collectors were already visiting Mapimí and collecting in the mines. In 1946, a huge pocket of adamite (zinc arsenate hydroxide, mostly occurring as yellow, green, or occasionally purple crystallized spheres, wheels, or balls) was found by Dan Mayers and Francis Wise (Moore and Megaw 2003:43). This discovery helped to spark interest in the locality, as did the presence of the U.S. mineral collector and dealer George Griffiths, who ran a store in Gómez Palacio (la Casa de las Rocas), and who arranged visits to Mapimí for visiting collectors from the United States. The store is still operated by Griffiths's widow, Catarina (Cata) Griffiths.

In the early days, miners and their wives exchanged specimens for food and clothing, but by the 1960s, the specimen market had become complex and highly commercialized. By this time, the practice of collectors traveling to Mexico either to self-collect or purchase specimens became sufficiently well established to warrant the publication of *A Field Guide to the Gems and Minerals of Mexico* with special emphasis on collecting opportunities (Johnson 1965). Rather than a scientific treatise, this was a guidebook to help those who wished to travel to mineral localities in Mexico to collect. In addition to mineralogical information, it included strategies for obtaining permits and information on how to purchase specimens in different localities.

In his description of Mapimí, the author advises:

> Upon arrival in Mapimí it will take about 15 *seconds* for the word to be spread that the *norteamericanos* have arrived. Then you will be deluged with friendly women and children selling *riscos* (specimens) from cardboard boxes. (Johnson 1965:29; emphasis in the original)

This description is clearly aimed at the relatively small-scale collector; richer and more serious collectors and dealers would arrange matters through Lázaro de Anda, George Griffiths, or, at a somewhat greater distance, dealers in El Paso and Tucson. As in Guanajuato and other localities, the mineral market quickly became organized hierarchically, with distinct tiers of sellers, buyers, and specimens.[8]

The cooperative ceased commercial mining in the early 1980s, but soon thereafter began selling specimens to Mike New of Tucson, Arizona. After a number of years of conflict with dealers in the nearby town of Bermejillo (the Dávila family) and others, New in 1995 secured a contract with Peñoles to extract specimens and formalized his contract with the cooperative, which now has fewer than ten members. The cooperative's work is subsidized by New, who then purchases minerals through Lázaro de Anda, the largest dealer in Mapimí and son of a former cooperative member and one of the town's first specimen dealers. New's company, Top-Gem Minerals, continues to hold the largest formal contract with miners and dealers in Mapimí (though others still buy directly from miners or from middlemen within Mapimí and from Chihuahua City and Ciudad Juárez) (Mike New, personal communication).

The Ojuela mine has been the site of some spectacular discoveries over the past few decades. The legrandite find of 1977, in which the two world-famous specimens the Aztec Club (now in the American Museum of Natural History) and the Aztec Sun were discovered, together with about twenty other magnificent legrandites, put Mapimí on the map for world mineral localities. Legrandite, a rare mineral that occurs in zinc- and arsenic-bearing deposits, was little known to collectors until the 1960s. The Tucson dealer Gene Schlepp recounts being shown legrandite specimens from Ojuela in 1962, saying, "he [a miner in Mapimí] wanted to trade a box of the stuff. This was unheard-of. We were buying it by the kilo. I didn't even know what it was." The best specimens from the 1977 discovery are now worth over US$1 million each.

Later, in October 1981, a pocket of adamite crystals with rare purple terminations was found in the San Judas vein, level 6. These specimens, after a series of attempted thefts and other shady dealings, were recovered by New and brought to the 1982 Tucson show, where they caused a sensation. As Mike New describes it, "the San Judas purple adamite specimens made their debut at the 1982 Tucson Show. Every one was exceptional . . . with the best cabinet specimen being priced at well over $10,000. One of the top pieces was purchased in 1993 by John Barlow: it is the 4×5.5×7-cm group pictured on p. 336 of *The F. John Barlow Mineral Collection*" (New, quoted in Moore and Megaw 2003). New's description of the movement of stones from fortuitous discovery to final enshrinement in famous magazine covers and collections is characteristic of specimen biographies as told by collectors and dealers. The passage addresses a like-minded audience of mineral lovers (those who know the specimens in their later lives), asking them to imagine the moment when these now famous specimens were first seen by human eyes.

The Ojuela mine is one of the world's most famous and productive mineral specimen localities, for several reasons: its mineralogical complexity; the fact that the mine continues to operate at a low, labor-intensive, and capital-extensive level, with a trained workforce in a context of high unemployment; and the density of Mapimí's social and economic ties to Tucson.

The National Museum of Natural History / Smithsonian Institution

My fourth field site differs from the other three, in that most of my research into it has been archival. The mineralogical collections of the National Museum of Natural History (part of the Smithsonian Institution) have played a central role in the development of the mineral sciences in the United States, and they also hold the material evidence of the country's expansion in territory, wealth, and power over the course of the nineteenth and twentieth centuries.

The Smithsonian Institution, which functions as an independent agency of the federal government, was established by a congressional act in 1846 and was intended to fill multiple aims: an observatory, a scientific research institute, a national library, a publishing house, and a museum. The funds for the Smithsonian came from an unlikely source: In 1828, James Smithson, a British naturalist and mineralogist, bequeathed his fortune (in the event of his nephew dying without heirs) "to the United States to found in Washington, under the name of the Smithsonian Institution, an establishment for the increase and diffusion of knowledge among men" (Smithsonian Institution Libraries 2012). In 1835, when Smithson's nephew died heirless, this rather extraordinary bequest to a country Smithson had never visited provoked a storm of discussion as to how best to fulfill the terms of the will.

In 1858, the natural history collections housed in the U.S. Patent Office were transferred to the Smithsonian's red sandstone castle on the Washington Mall. These exhibits consisted largely of artifacts from the Wilkes United States Exploring Expedition of 1838–1842 and other expeditions, and the old "patent model exhibit" illustrating American inventions (Henson 1995:S251–252).[9]

The Smithsonian's first secretary, Joseph Henry, considered investing in a museum a poor use of the limited funds of the bequest, because such efforts would only reach a limited and local audience. However, his successor, Spencer Baird, took a different view, and dedicated much of his time to cultivating and promoting the National Museum. His efforts included accessioning specimens from the U.S.-Mexican Boundary Survey and other surveys in the expanding West to the Smithsonian collections. He hoped that the sheer mass of material would eventually make the establishment and support of a national museum unavoidable.

Baird's great triumph came in the 1870s, while still assistant secretary, when he was appointed to head the interagency committee to develop the exhibits for the Centennial Exhibition of 1876 at Philadelphia. At the close of this massively popular exhibit, visited by nearly 10 million people or one-fifth of the nation's population (Rydell 1987:10), Baird convinced many of the exhibitors to donate their exhibits to the National Museum (Henson 1999:S253). These donations, filling some sixty railroad cars, formed the core of the national collections, and necessitated the construction of a new building, the Arts and Industries Building, to house them. Beginning in 1881, the accessions were moved to the Arts and Industries Building, and a National Museum was thereby founded. In 1911, a new building was constructed on the other side of the National Mall; it is this building that now holds the National Museum of Natural History.

Mineralogical and geological collecting began in the institution's very first years.[10] The collections of the Wilkes United States Exploring Expedition to the South Seas (1838–1842) included mineralogical and geological specimens gathered by the expedition's mineralogist, the eminent scientist James D. Dana, who later founded the Dana system of mineralogy by which minerals are classified (with some modifications) even today. These collections went to the U.S. Patent Office, where they were stored until the formation of the National Museum in 1881. The mineralogical and geological collections of the 1876 exhibition were extensive and won many awards from the exhibition's judges; they too were integrated into the National Museum.

Over the last quarter of the nineteenth century, the institution continued to collect and curate mineralogical and geological specimens from different parts of the country and to send exhibits to the various exhibitions taking place during this time. The expanding surveys of the western United States and the activities of the U.S. Geological Survey, which was founded in 1879, also fueled the collections tremendously. By 1911, when the museum took up residence in its new building, the economic geology exhibits, which included ores, rocks, and ornamental and building stones, were separated from the so-called systematic collection, which included minerals and gems (figure 1.6). The systematic collection followed the Dana system and tended to present minerals as expressions of the laws of nature rather than as examples of natural resources, though in an expanding U.S. empire, the two were closely connected.[11]

During these years, the collection's growth was at the mercy of outside circumstance and the whim of donors, for curators had few funds either for purchase or field research. F. W. Clarke, the curator for the Department of Minerals, wrote in the 1884–1885 Annual Report:

> Good as the collection is in some directions, it is marred by many gaps,
> and has to contend with serious difficulties. Its gain has been mainly
> through gifts and exchanges; whereas other museums have funds
> for the purchase of specimens, and can build up their series to better
> advantage and with greater scientific system. We . . . are handicapped

FIGURE 1.6. Smithsonian mineral exhibits, National Museum of Natural History, 1919. Smithsonian Institution Archives. Reprinted with permission.

in our competition with others, and are compelled to develop too much at haphazard. The one thing our department of minerals needs is a fund for purchasing material and until such a fund is supplied we must remain at a serious disadvantage. Under our present system we can develop a good working collection, but a *great* collection, a *model* collection, is beyond our reach. (CAR 1884–1885)

This situation continued, more or less, until the banner year of 1926, when the museum received two major bequests, respectively, from the estates of Frederick Canfield and Washington Roebling, both premier mineral collectors. The Roebling bequest contained sixteen thousand specimens and included a US$150,000 endowment for specimen purchase. The Canfield collection, which, in the words of the curator "came as a pleasing surprise, no intimation having been received that Mr. Canfield had considered the Smithsonian as a depository" (CAR 1926–1927), contained nine thousand specimens and was accompanied by a US$50,000 endowment. These bequests and funds enabled the Smithsonian mineral collection, over the succeeding decades, to become among the very finest world mineral collections.

The Roebling and Canfield collections included multiple specimens from outside the United States, and the funds allowed for the purchase of still more specimens, including foreign ones. This put the national collection on a global scale, making it comparable to collections at Harvard and the Philadelphia Academy of Sciences and to the museums of natural history in Paris, London, and Vienna.

The Mexican minerals in the collection began to grow in a serious way in the 1920s, when the curator, William Foshag, began his fieldwork in Mexico. During his first few trips, he was purely engaged in research for the museum and in collecting, but in 1942–1946, he worked directly for the U.S. War Department. In these trips, he spent most of his time traveling through Mexico investigating sources of borate minerals, which could be used to produce high-energy jet and rocket fuel. Foshag's experience in Mexico, his close ties to Mexican scientists, and his earlier research into borate minerals in California made him the ideal person for this job. In the process, he brought back specimens (purchased with Roebling funds) for the national collections. For instance, the *Curator's Annual Report* for 1943–1944 states that the following Mexican specimens were obtained with money from the Roebling Fund:

> a collection of 70 specimens of unusual calcite crystals with a new crystal habit from La Aurora mine, Chihuahua, Mexico; a series of 8 minerals from Mexico purchased by W. F. Foshag, consisting of 2 crystallized argentites (silver sulphide) from Guanajuato, 2 native silvers from Bato-pilas, 1 legrandite (zinc arsenate) from Coahuila [legrandite does not occur in Coahuila; this specimen probably came from the Ojuela mine in nearby Durango], 1 aguilarite (silver sulphide and selenite) from Taxco; 1 stephanite and 1 pyrargyrite (both silver antimony sulphides) each from Zacatecas and 1 specimen of natrolite from Livingston, Montana. (CAR 1943–1944)

Foshag also worked with Mexican geologists in the observation and analysis of the 1943 birth and eruption of the Paricutín Volcano in Michoacán, Mexico, and deposited samples, photographs, and reports in the museum's collections and the Smithsonian Institution Archives. Although Mexico per se has no special classificatory status within the collections, which are organized by mineral type and locality but not by nation, and although the collections are dedicated to comprehensiveness and do not particularly focus on any one area (at least outside the United States), Foshag's interest in Mexico certainly helped those areas of the collection to improve. In addition, the peculiar relationship between the two nations, as well as their geographic nearness, increased the emphasis on Mexican minerals.

Interest in Mexico at the museum became less focused after Foshag's time. However, in Desautels's early years at the Smithsonian, he too conducted mineralogical research into Mexican minerals, including a study with Roy Clarke on legrandite from Mapimí (Desautels and Clarke 1963). Under his direction, the museum continued to acquire some of the best minerals coming out of Mexico. In a letter to his superiors at the Smithsonian justifying a collecting trip in Texas and Arizona, Desautels wrote:

> Since Mr. Foshag's death, our contacts with Mexican sources have been weak at best. Mexico is currently producing more new and fine specimens than any other country and we are obliged to obtain the best of them. . . . On previous trips to the Southwest I have obtained verbal agreements from all the important wholesalers to consider our needs first. On this trip I was able to extend the connections down to several sources in Mexico. (Paul E. Desautels's papers, letter dated April 6, 1964)

One of these sources in Mexico was the Chihuahua dealer Benny Fenn, who in 1970–1971 sold a six-hundred-piece collection to the museum that "added greatly to what was already the finest and most complete suite of Mexican minerals ever assembled" (CAR 1970–1971). The collections described in chapter 4—particularly the Harvard Mineralogical Collections and the Miguel Romero Collection—contest this claim, but the statement underscores the comfortable place held by Mexican minerals in the collections. This period also coincided with the increasing flows of Mexican minerals coming through Tucson to the rest of the world.

As the years went by, Desautels turned away from mineralogical research and toward collection building, and he helped to develop new forms of appreciation for minerals not based on locality but on aesthetics.

2 SHIFTING STONES: MINERALOGY AND MINERAL COLLECTING IN MEXICO AND THE UNITED STATES

Value—the process of rating things as meaningfully different—only works as a generative, dynamic force because it is embedded within a similar process at a different categorical scale, where people judge the criteria by which objects *should be judged* to differ in significant ways—such as privileging rarity as a valuable quality for minerals. Change at this level—making difference meaningful—happens more slowly and is harder to see, but it is nevertheless an indispensable aspect of value-making.

Two major shifts can be identified in the valuing of minerals in Mexico and the United States that shed light on this second level, in which the kinds of difference that make a difference are stabilized and destabilized. Over the course of the nineteenth century, the perceived center of mineralogical expertise shifted from Mexico to the United States, a shift that was somewhat but not entirely rebalanced in the twentieth century. In the past thirty years, a second shift has occurred with the rise in popularity of so-called aesthetic minerals in the United States and Europe (and to a lesser extent, Mexico). This trend has successfully installed *aesthetics* as a primary valued quality for minerals, based on a comparison between minerals and fine art.

Mineralogy—Mexico and the United States

The word *mineral,* imported from the Celtic, entered Latin in the twelfth century and was used to designate metallic ores, in contrast to *lapis,* which referred to rocks and silicates. The first use of the word *mineralogy* came in 1690, in Robert Boyle's essay "A Previous Hydrostatical Way of Estimating Ores." Extended forays into the scientific examination of the physical, chemical, and crystallographic properties of inorganic matter did not happen until the late eighteenth century, with the work of Antoine-Laurent de Lavoisier and his students in chemistry and René Just Haüy in crystallography (Greene and Burke 1978:5). In the last two decades of the century, the study of minerals located itself in the state-sponsored schools of mines that had been established over the previous century in Europe. These included the Mining Academy at Selmecbánya in the kingdom of Hungary and the Bergakademie at Freiberg in Saxony, where Abraham Gottlob Werner served on the faculty from 1775 until 1817. Werner's two main claims to fame are his establishment of the world's first system of mineral classification (based on the external physical characteristics of minerals) and his position as a leading proponent of Neptunism—the theory that the major planetary building block basalt had formed out of a planetary ocean saturated with minerals (Greene and Burke 1978:6).[1]

In their comprehensive article on early mineralogy in the United States, John C. Greene and John G. Burke write that:

> The influence of these schools [the mining schools of continental Europe] on the development of mineralogical science can scarcely be exaggerated. Their success is demonstrated by the rapid rise in the production of minerals in Europe in the nineteenth century and by the exploitation of the mineral resources of European colonial empires around the globe. Their combined practical and theoretical curricula produced graduates who made important contributions both to mining technology and metallurgical practices and to the sciences of geology, mineralogy, and crystallography. (1978:7)

The Science of Minerals in Nineteenth-Century United States and Mexico

This model of a state-sponsored school of mines was continued in New Spain (colonial Mexico and Guatemala) with the founding of the Real Seminario de Minería in Mexico City on January 1, 1792 (Uribe Salas and Cortés Zavala 2006:493–494). The school, the first of its kind in the New World, was founded as part of the Spanish crown's efforts to revitalize and modernize mining in its colonial possessions, especially in New Spain. The Real Seminario's first director, Fausto de Elyuhar (discoverer, along with his brother Juan José, of the element tungsten), had studied at the Mining Academy at Selmecbánya. He recognized that the mining industry suffered from a lack of knowledge of the geology and mineralogy of the region as well as a paucity of technological expertise (Uribe Salas and Cortés Zavala 2006:494). The Real Seminario, which became the Colegio de Minería upon Mexico's independence from Spain, trained generations of scientists and engineers who populated schools of mines and government agencies all over the country (Uribe Salas 2006).

Accordingly, in 1795, the school inaugurated the first course in mineralogy in the Americas, to be taught by the Spaniard Andrés Manuel Del Río (figure 2.1). Del Río was a Spanish mineralogist who had trained at the Escuela de Minas in Almadén, Spain, the École Royale de Mines in Paris, and the Bergakademie in Freiberg, Saxony. One of his fellow students at Freiberg, who would later visit and collaborate with him in New Spain, was Alexander von Humboldt (Ramírez 1891; Uribe Salas 2006:239). After completing his schooling in Freiberg, Del Río went on to study in Shemnitz, Hungary, and to tour mining facilities in Bohemia and Austria. He studied chemistry in Paris, but he was forced to leave (disguised as a water-bearer) in the wake of the French Revolution, after his mentor Antoine-Laurent de Lavoisier was guillotined on August 8, 1794 (Uribe Salas 2006:242). He went to England, Austria, Transylvania, and Saxony. He ended up in Vienna, where he received notification from the Spanish government that he was to be sent to New Spain to aid in the formation of a school of mines in Mexico City. He spent nearly all the rest of his life in the Americas, all but six years of it in Mexico.

Del Río's work in Mexico combined teaching, translation, writing, research, and practical fieldwork. Until his death in 1849, with some breaks, he taught at the Real Seminario, which was renamed El Colegio de Minería after Mexican Independence in 1810. His pupils went on to become some of the most prominent mineralogists and mining engineers in nineteenth-century Mexico. As part of his labors training new generations of mineralogists, Del Río drew on his skill as a polyglot, translating the works of Werner, Karsten, and others for use in the Real Seminario. In addition, he wrote his own multivolume work, *Elementos de la Orictognosia* (Werner's term for "practical" mineralogy) for use by his students. This textbook was regarded by Humboldt and others as the best mineralogical work of its time in Spanish (Arnáiz y Freg 1948:17).[2]

In 1821, when Mexico finally won the War of Independence from Spain, Del Río was invited to remain in Madrid, but he chose to return to Mexico. In 1829, the Mexican government expelled Spaniards from Mexico, including Fausto de Elhuyar. Though he was granted an exception, Del Río also chose to leave and went to Philadelphia from 1829 to 1834. There he became a member of the American Philosophical Society (the oldest scientific society in the Americas), one of the earliest centers for

FIGURE 2.1. Andrés Manuel Del Río.

science in the U.S. republic. He returned to Mexico in 1834 and remained there until his death in 1849. He was a member of scientific societies in Mexico, the United States, Spain, France, and Germany.

The Colegio de Minería operated for decades as the primary mining academy of the Americas. It trained generations of mining engineers, geologists, mineralogists, topographers, and other professionals who peopled the mines, universities, and government agencies of nineteenth-century Mexico. They included Casimiro Chowell, who headed the insurgent forces in Guanajuato against the Spanish in the War of Independence and was hanged in front of the Guanajuato city granary, the Alhóndiga de Granaditas; José Salazar Ylarreguí, Mexican commissioner of the U.S.-Mexican boundary survey in the 1850s (after the death of Pedro García Conde; see chapter 3); and Lucas Alamán, historian of Mexico and leading Conservative politician.

In contrast to the promising atmosphere for the earth sciences in early nineteenth-century Mexico, mineralogy in the United States during this time was a topic for parlor scientific societies, mostly in Philadelphia, Pennsylvania; New Haven, Connecticut; and Cambridge, Massachusetts. There were few institutions or publications devoted to the study of mineralogy or of the sciences more generally. Greene and Burke write:

> Before the American [U.S.] mineralogists lay a vast continent whose mineral resources invited discovery, description, and exploitation. But . . . the hard fact was that American society had no ready replacement for the European system of patronage by government subsidy and the private munificence of a leisure class. The British models on which the American governments were formed afforded no precedent for extensive governmental support of science and industry, and the Jeffersonian principles which soon prevailed in American politics reduced still further the prospect of public patronage for either. Everything—scientific societies, libraries, collections, colleges, and universities—had to be built from scant beginnings. In these circumstances it is not surprising that the science of mineralogy acquired a solid footing in the United States only gradually, remaining on the descriptive level well into the nineteenth century (1978:20)

The disparity in institutional support for the earth sciences between the United States and Mexico derives from two sources in particular: the fact that England did not have a school of mines comparable to those on the continent and, therefore, neither funded nor provided a model for such a school in the New World; and the lack of a significant mining economy up until the Gold Rush of 1849. The early years of the century were taken up with establishing the teaching of mineralogy in universities and in field research to help describe the geology and mineralogy of the expanding United States.

In 1816, Parker Cleaveland published his mineralogy text, *An Elementary Treatise on Mineralogy and Geology*, which was adopted by many college courses in mineralogy, as these began to be established in the first decades of the nineteenth century (Wilson 2012:Parker Cleaveland, 1780–1858). In 1837, J. D. Dana, a professor of geology at Yale, published the first edition of his *System of Mineralogy*, which described the 352 species known at the time, borrowing from biology and botany the concept of species and a basic system of classification (Dana 1837; Gray 2002). It was not until the fourth edition (1854) that Dana devised the system based on the chemical composition of minerals[3] that, with some modifications, is used today (Gray 2002; Newell 1997). The publication and subsequent acceptance of the Dana system marks a watershed in the prestige of U.S. mineralogy.

The growth of U.S. mineralogy as a science "of international stature" (Greene and Burke 1978:107) was embedded within larger political and economic shifts. The most obvious of these were the period of intensive westward expansion that began after the Civil War, the building of the railroads, and the dramatic exploitation of silver and copper in the 1880s and 1890s in Nevada, Arizona, and California, in particular. Beginning in the 1890s (after the passage of a more favorable mining law, from the point of view of foreign investment), U.S. companies began moving aggressively into northern Mexico. They brought electric power and cyanide processing to many old mining localities, opened new mines, and established smelters and company railroads. For instance, between 1897 and 1913, about seventy mining companies (most were U.S.-owned) operated in Guanajuato (Meyer Cosío 1999:101). In 1910, on the eve of the Mexican Revolution, the Guggenheim brothers' investments in Mexico totaled US$50 million, making the American Smelting and Refining Company, or ASARCO, the largest private corporation in Mexico at that time (Bernstein 1965:50–56; Durnett 2009).

Along with the new importance of mining in the United States came governmental and private investment in the study and teaching of the earth sciences. In 1864, the School of Mines at Columbia University was founded; the Colorado School of Mines followed in 1874.

In 1879, the U.S. Geological Survey was founded to support fieldwork and research of the geology and mineralogy of the whole nation (individual states had been setting up geological surveys since the 1830s).

TABLE 2.I. SIGNIFICANT MOMENTS IN THE EXPANSION OF
THE U.S. MINING ECONOMY AND IN THE GROWTH OF THE
EARTH SCIENCES IN THE UNITED STATES (1848–1890s)

1848	Treaty of Guadalupe Hidalgo ends Mexican-American War. Mexico loses over 50 percent of territory to United States
1849	start of the California Gold Rush
1854	4th edition of J. D. Dana's *System of Mineralogy*, which proposes a new system of classification, based on chemical composition
1864	founding of the Columbia School of Mines
1863–1869	building of the first transcontinental railroad
1874	founding of the Colorado School of Mines
1876	William P. Blake, on behalf of the Smithsonian, solicits contributions of rock and mineral collections, by state, for the Centennial Exhibition at Philadelphia. The majority of these are donated to the U.S. National Museum at the close of the exhibition
1879	founding of the U.S. Geological Survey
1870s–1890s	silver and copper booms in Arizona
1884	new mining code in Mexico to promote foreign investment
1888	establishment of the Geological Society of America
1890s	movement of U.S. mining companies into Mexico

Table 2.1 shows how mining expansion and growth of the earth sciences in the United States overlapped in time. Subsoil resources provided the warrant (and the capital) for institutions dedicated to the earth sciences.

During this same period, Mexican mineralogy was, on balance, losing ground.

The War of Independence from Spain (1810–1821) and the subsequent political instability (between 1825 and 1855 the presidency changed hands forty-eight times) made the establishment of institutions capable of promoting scientific research exceedingly difficult. José Alfredo Uribe Salas and María Teresa Cortés Zavala write:

> The uncertainty and political instability that characterized Mexico after the War of Independence depleted attempts by the different governments to re-establish the educational system on the basis of support for the cultivation of the sciences, so that activities related to geology were confined to a battered Colegio de Minería and a few foreign geological explorers. (2006:500)

The political turmoil of mid-nineteenth-century Mexico took its toll on the Colegio de Minería. For instance, in January 1842, a note in the college's records states blandly that "the Director asks for authorization to pay 467 pesos 3 reales to repair the damage to the building caused by cannon and gunfire in the Revolution in September" (Ramírez 1890:305), that is, General Santa Anna's coup against the Mexican presidency, which was victorious on October 10, 1841.

The accompanying decline in the mining industry, in terms both of profits and political power, also affected the Colegio de Minería's fortunes. Two educational reforms from above forced the college to alter its curriculum, moving away from an emphasis on mining and earth sciences. In 1867, the college was closed and reopened as the National School of Engineers. The failed efforts of the Mexican Boundary Commission (whose job it was to survey the new U.S.-Mexican border after the Mexican-American War) to receive personnel or equipment for making scientific collections (see chapter 3) provide a case in point of this mid-century decline in Mexican mineralogy's fortunes. At the same time, the state did manage to carry out three scientific/cartographic missions, of the Isthmus of Tehuantepec (1823–1826), of the area surrounding Mexico City (1825), and of the northeastern part of the border with the United States (1827–1831) (Azuela 2009:103). Also, literary and scientific institutions were established by the governments of several different states (these became the state universities) that taught earth sciences along with botany, biology, and other fields (Uribe Salas and Cortés Zavala 2006:498).

In the latter part of the nineteenth century, Mexican earth scientists were able to consolidate educational and research institutions and surveys such as the Sociedad Científica "Antonio Alzate," founded in 1884, and the Instituto Geológico de México,[4] founded in 1888 (it had been established two years earlier as the Comisión Geológico de México, dedicated to the investigation of Mexico's natural resources). One of the Instituto Geológico de México's founders, José Guadalupe Aguilera, a graduate of the Colegio de Minería, traveled through Europe visiting expositions and museums for inspiration and to forge connections. These institutional and intellectual labors laid the groundwork for a resurgence of the newly professionalized earth sciences in Mexico in the last quarter of the century (Morelos Rodríguez 2010).

Siglo XX / Twentieth Century: U.S.-Mexican Collaboration and Competition

The twentieth century began strongly for Mexican mineralogy, with the X Congrés Internationelle Géologique held in Mexico City in 1906.

The magnificent new building of the Instituto Geológico de México, located in the neighborhood of Santa María la Ribera in what was then the outskirts of Mexico City, was finished in time to host the congress (Irazaba-Ávila and Espinosa Arrubarrena 2002). This was the first time that the congress had been held in the Americas, and it was a great occasion, with several days of presentations and field trips to mining localities and sites of geological interest within easy reach of Mexico City. As part of the support for these trips, thirty-one guidebooks were written on the geology and mineralogy of different regions (De Cserna 1990:13). The congress participants brought mineralogical, petrological, and paleontological collections with them, many of which remain in the current Museum of Geology, housed in the institute's old building.

The institute also worked to build its national collections, soliciting contributions from different parts of the Mexican republic. For instance, on June 25, 1903, the institute's assistant director, Ezequiel Ordóñez, wrote to Ponciano Aguilar, a mining engineer and prominent citizen of Guanajuato (see chapters 3 and 4), to solicit samples of local building stones for a national collection of "construction materials." Later that year, Aguilera wrote to remind Aguilar and to ask for a specimen of the mineral aguilarite, discovered in 1891 and named for Aguilar, for the institute's mineral collection (Ordóñez 1903). In 1904, Ordóñez, along with a number of foreign geologists, located the first economically viable oil well in Mexico, ushering in Mexico's age of oil (De Cserna 1990:16). The institute's activities declined between 1910 and 1929 (the period of the Mexican Revolution and the tumultuous years following), when it was reformed as part of the Universidad Nacional Autónoma de México (UNAM) under the name Instituto de Geología. In 1956, the Ciudad Universitaria was established in the southern part of the city, and the institute was transferred to the new campus. The 1906 building has remained as a museum.[5]

In the 1920s, collaborative work among Mexican geologists and mining engineers and the Smithsonian Institution began, primarily through the activities of Smithsonian curator of mineralogy William Foshag (figure 2.2). In 1927, Foshag took the first of his many trips to Mexico for the purpose of conducting field research and collecting mineral and rock specimens for the National Museum, with Harry Berman from Harvard University as his research assistant. In the publication *Explorations and Field-Work of the Smithsonian Institution in 1928,* Foshag reported that

FIGURE 2.2. William Foshag in Mexico. Smithsonian Institution Archives. Reprinted with permission.

they first went to Mexico City to arrange for official permits and to meet with scientists at the Instituto Geológico, and then visited Guanajuato, of which he said, "the rich bonanza ores of the early days now appear to be entirely exhausted, but mining is made profitable by the efficient treatment of large bodies of low grade ore" (Foshag 1928:21). They then traveled to mining districts in Zacatecas and from there to the Ojuela Mine in Mapimí, Durango, where Foshag "was able to collect, together with pyromorphite and other rare and beautiful minerals, a number of good specimens of the very rare species carminite, an arsenate of lead and iron" (Foshag 1928:23).

The team then continued to Naica, Chihuahua, where, with the permission of Peñoles, the mining company, they extracted the two groups of gypsum crystals now in the National Museum of Natural History and the Harvard Mineralogical Museum (see chapter 3). Finally they visited Santa Eulalia and Los Lamentos, also in Chihuahua. Foshag ends the article with gracious acknowledgment of "a number of the mining men in Mexico, without whose active cooperation little could have been accomplished" (Foshag 1928:26).

This trip was the first of a number of similar research and collecting trips. In the early years, these trips were conducted solely on behalf of the Smithsonian, but beginning in 1940, Foshag was contracted by the U.S. Geological Service in collaboration with the War Department to investigate possible strategic uses of minerals in Mexico. This work kept him in Mexico for much of the following decade. During that time, in addition to exploring the potential uses and sources of tin, antimony, and manganese, Foshag also continued to cultivate relations with Mexican scientists—particularly at the Instituto Geológico—and to collect minerals.

Foshag became one of the primary investigators of the extraordinary emergence of a volcano on February 20, 1943, in the town of Paricutín, Michoacán. The volcano literally grew out of a cornfield, reaching a height of 150 meters in six days. Over the course of the first two years of its life, it buried the town of Paricutín and much of the neighboring town of San Juan Parangaricútaro. The volcano erupted several times over the course of the next decade and returned to dormancy in 1952. It provided an exceptional opportunity to trace the life span of a new volcano (the newest in the Western hemisphere and the most rapidly growing ever known).[6] In an article in *Time* magazine in 1994, Foshag was quoted as saying, rather breathlessly, "It is, I believe, just as spectacular as Vesuvius ever was, and in its more violent phases it is better" (*Time* 1944).

In studying the volcano from birth to dormancy, Foshag was joined by Mexican geologist Jenaro González Reyna (of the Instituto de Geología) and by a local resident, Celedonio Gutierrez, who became an exceptionally observant and capable field assistant. Foshag and González

Reyna published their study of the volcano in a report for the U.S. Geological Survey in 1956 (Foshag and González Reyna 1956).

During his years of work in Mexico and on topics in Mexican mineralogy, geology, and volcanology, Foshag cultivated close working relationships with Mexican scientists, mining engineers, and mineral and antiquuities dealers. His friendship and working partnership with Dr. González Reyna was especially close. In 1949, after the two had been working for many years, Foshag nominated González for fellowship in the Geological Society of America, stating that "I have known Sr. González for most of his professional life as a geologist, and have worked in collaboration with him during the war years on the survey of mineral resources of Mexico" (SIA Record Unit 7281, box 2, letter to Henry Aldrich, Geological Society of America, March 16, 1949).[7]

In 1956, just a few months before Foshag's early death, perhaps from emphysema, González wrote to thank him, with great affection, for naming one of the minerals found in the sublimates of Paricutín after him, gonzalezite, along with another mineral, paricutinite. (These were in fact never registered as new minerals, perhaps because Foshag did not have time to publish his description before he died.) González also asked for specimens of each mineral for the museums of the Instituto Geológico and the University of Guanajuato, his alma mater. He also promised to send a specimen of granodiorite from the copper mines of El Marqués, Michoacán. He also reported that he had not yet received the copies of the bulletin they had published on the first three years of the Paricutín volcano, but that he hoped to get them soon. He closed the letter with a conventional but very warm valediction, saying, *"Con saludos muy afectuosos de nuestra parte, sabes que somos como siempre sus amigos* (with very affectionate greetings from us [his wife Pachita, their children, and himself], you know that we are as always your friends)" (SIA Record Unit 7281, box 2, letter to William F. Foshag from Jenaro González Reyna, January 16, 1956). Thus the multiple transactions of letters, specimens, geological publications, new mineral names, and so on helped to build a strong professional and personal relationship, based on commonality of purpose.

At the same time, in collecting mineralogical samples and publishing reports for the Smithsonian, Foshag and González helped to increase the value of U.S. mineralogical research and collections at the expense of Mexican research and collections. The removal of the Naica gypsum crystals is a case in point, as we will see in chapter 3. The two best specimens went to the Smithsonian and Harvard. The Museo de Geología in Santa María la Ribera does have a gypsum group from Naica, but it is only about a quarter the size of those in Washington and Cambridge.

Also, in spite of his fruitful collaborations with González, Foshag's opinion of geology and mineralogy in Mexico was more muted. For instance, in 1955, he responded to a request for information about studying geology in Mexico from Dr. William H. Parsons of Wayne University in Detroit, saying:

> I can understand why your student would be interested in studying in Mexico. Life there is, as they say, "suave." But I think that after he has finished his course with you, the University of Mexico [UNAM] or the Polytechnic University would offer him little more in the academic line. If, however, he expects to pursue geology in Mexico, he could get a good background in the details of Mexican geology. . . . I am afraid that I cannot recommend the University of Mexico for post graduate training, but only as a happy experience. (SIA Record Unit 7281, box 2, letter to William H. Parsons, December 16, 1955)

Foshag's remarks here put mid-twentieth-century Mexican earth science "in its place," as it were, locating it as a particularist form of knowledge, good for learning about "Mexican geology" but not geology as a whole.

In the second half of the twentieth century, distinguished work in descriptive mineralogy and crystallography was done by Eduardo Schmitter, professor of mineralogy and petrology at the UNAM, and Francisco Fabregat, professor of crystallography, also at the UNAM. In 1984, the Sociedad Mexicana de Mineralogía was founded and has held periodic conferences, including in September 2011 (Rubinovich 1993; Alfredo Victoria Morales, personal communication). Research in the earth sciences continues at the Instituto de Geología of the UNAM and in other schools around the country, and new efforts have begun to focus on the preservation and use of historical mineralogical collections in Mexico City, Guanajuato, and elsewhere.

The case of geology and mineralogy in nineteenth- and twentieth-century Mexico shows the effects of a shift in making difference meaningful. Once Mexican mineralogy loses its international stature, it becomes that much harder to keep good collections in the country, to get funds for geological exploration, to attract international students (except those in search of a "happy experience"), and so on. A new landscape of value is stabilized that then makes it difficult for the scientific work done in Mexico to attract international attention.

Mineral Collecting—Curiosities and Specimens

Rocks and minerals formed part of many of the early cabinets of curiosity of the sixteenth and seventeenth centuries, but in keeping with the eclectic logic of these proto-collections, they were usually collected as part of a more general category of *naturalia,* including zoological and botanical specimens, and often in a cabinet that also contained *artificialia,* or objects made by humans. Only a few collectors, usually connected to mining or medicine (the two areas where minerals figured most prominently), focused particularly on minerals; of these, the most famous was Georgius Agricola, author of *De Re Metallica,* the first treatise on mining (Wilson 1994:19–21). For the most part, minerals formed parts of cabinets of curiosities as compendia of the natural and sometimes also the human-made world.

As the name suggests, cabinets of curiosity made a virtue (a value, in the terms of this book) of the curious. The designation *curious* tended to refer to objects, possibly from far away or long ago but at any rate alien in some way to the collector and viewer, that excite interest on the basis of novelty or peculiarity. In some cases, *curiosity* (in the sense of the quality of being curious) was immediately apprehensible, but in others a degree of connoisseurship was needed to appreciate such a cabinet properly (Pomian 1990). *Curiosity* as a valued quality emerged in the context of the exploratory voyages of the early period of colonial expansion, was often applied to ethnological specimens, and was sometimes explicitly contrasted to *beauty* (Thomas 1991). More accurately, perhaps, to call an object "curious" was to render the question of its beauty irrelevant.

Over the course of the period from 1500 to 1800, roughly speaking, several things happened to cabinets of curiosity: They became more systematically organized; they moved from emphasizing "marvels" to concentrating on the collection of typical forms; and they became more democratic, moving from an esoteric province of kings and aristocrats to the founding collections of museums aimed at a bourgeois public (Bennett 1995; Daston and Park 1998; Findlen 1995; Impey and MacGregor 1985; Pomian 1990).[8]

Increasing systematization also meant that the boundaries between the natural and the artificial became more rigid and more strictly enforced in the practice of collecting, so that natural and human-made objects were not usually present in the same collection (though ethnological specimens, as material signs of the "savage slot," that category between nature and society [Trouillot 1991], did sometimes share space with animals, plants, and minerals). Moreover, collections during this period tended to become more specialized, focusing only on minerals, plants, or animals, and within any one of these, only on particular geographic origins or other more specific criteria.

As the science of mineralogy developed, beginning in the eighteenth century, mineral collections became both a source of data and a didactic tool meant to instruct viewers about the mineral kingdom. René Just Haüy (1743–1822), the French crystallographer, and Abraham Gottlob Werner (1749–1817), the German mineralogist and primary advocate of Neptunism (the theory that basalt is of aqueous origin), both had important mineral collections. The state-sponsored mining schools founded in Europe in the eighteenth century also possessed mineral collections, as did many of the professors and students in these schools.

As these institutional and private collections consolidated, so did new values for the objects within them. Minerals, like other natural history objects, were valued for their ability to represent either the laws of science or the particular features of a given region. They were thus valued not for their peculiarity but for their typicality (at least with reference to whatever they were meant to depict or manifest). In the case of rocks and minerals (in some contrast to plants and animals), collections might also represent the natural resources and mineral potential of a given region or nation (Ferry 2010; Pomian 1990). The word *specimen*, which became the standard way to refer to objects in these kinds of collections, shows the objects' intended purpose. A *specimen*, according to the *Oxford English Dictionary*, is "an animal, plant, or mineral, a part or portion of some substance or organism, etc., serving as an example of the thing in question for purposes of investigation or scientific study." As collections became the heart of public museums, which are institutions with an inherent didactic mission, this ability to exemplify became even more valuable. This is not to say that curiosity, beauty, and other qualities played no part in the collection and display of minerals, but none of these was the quality valued most. That place was held by what we might call *exemplarity*.[9]

By the nineteenth century, a market for mineral specimens and a newly important profession, that of the mineral dealer, had emerged. Mineral dealers sold to both individuals and institutions. The archives of the British Museum of Natural History include invoices, correspondence, and catalogs from British and continental dealers dating from the first years of the nineteenth century. This became a primary source of minerals for museums, but they also employed and contracted field collectors to gather specimens on their behalf (Mueggler 2011). These sources allowed museums and private collectors to broaden their scope and to exercise more agency in the development of their collections, because they were no longer solely dependent on donations, government expeditions, and trades.

In the new republics of the United States and Mexico, the traditions of European mineral collecting were incorporated slowly. In the United States, as in England, collecting priorities were largely set by private collectors up until the middle of the nineteenth century. Moreover, because there were few established museums in the United States during these years, most of these collections have not survived, in contrast to the great nineteenth-century British collections, many of which are now in the British Museum of Natural History. Exceptions include the collection of Colonel George Gibbs, now at Yale's Peabody Museum, and the collection of Joel Roberts Poinsett, which was held for many years in the Philadelphia Academy of Natural Sciences and, upon the sale of that collection in 2006, went partly to the Smithsonian and partly into private hands.

The Centennial Exhibition of 1876 occasioned a spate of field collecting on the part of the Smithsonian Institution, which was charged with putting together state rock and mineral collections. The Smithsonian sent out circulars to each state government and hired a field collector, Thomas Donaldson, to travel to each state. These collections later went to the Smithsonian and, along with zoological, botanical, and ethnological material, formed the bulk of the National Museum of Natural History (Ferry 2010). Also during this period, natural history specimen dealers came to prominence in the United States. Their activities significantly affected the development of museums, research, and scientific education in the United States (Barrow 2000; Kohlstedt 1980). The main mineral dealers of the time, such as A. E. Foote and George English, hired agents to go to places like Mexico and collect or purchase specimens. Some collectors in Mexico, like Ponciano Aguilar in Guanajuato, also purchased specimens from U.S. and European collectors.[10]

By the turn of the twentieth century, mineral collecting had gained momentum in the United States, with many of the most important collectors being industrialists, mining engineers, and other powerful men. These included J. Pierpont Morgan, Washington Roebling (who designed the Brooklyn Bridge), Albert Holden (who owned numerous mines and smelters in the western United States and northern Mexico), and Frederick Canfield (a mining engineer who worked in zinc mines in New Jersey and silver mines at Potosí, Bolivia). Many of these collections were bequeathed to museums such as the Smithsonian and the Harvard Mineralogical Museum.

By the middle of the twentieth century, the hobby had become more widespread and democratic. Field collecting (collecting specimens oneself, rather than buying them) became a popular hobby for many young boys and also grown men. Mineral clubs and gem and mineral fairs sprang up in all regions of the country, with healthy membership numbers and robust attendance. Rock shops, small stores specializing in the minerals and rocks of the immediate region, were started in the 1950s as part of the emerging postwar car culture of the United States. Many of these were located on Route 66 and similar highways. These shops have largely died out, though a few still exist. A post on the Rockhounds listserv captures a feeling of nostalgia for these shops and the kind of mineral collecting they represented:

> Well kids, I have FINALLY found a real rock shop! I knew it was real right off the bat when that glorious smell of rock-saw lubricant greeted me as I walked in the door. Piles of rough in the back room, lots of cut material (oh my, the Mexican opals . . . sorry about the drool on the counter), and many MANY reasonably priced specimens (plus a lot of killer pieces at appropriate prices!). . . . It's not your foo-foo "new age" place. It's a shop like I remember from the Good Old Days. . . . Oh joy! Oh rapture! etc etc. (Brown 2005)

Many of the current mineral dealers and serious collectors, including those at the high end, got their start during these years field collecting in abandoned mines, quarries, and construction sites. However, most people I spoke with (both high-end collectors and dealers and the more populist rockhounds) reported that stricter trespassing laws and liability concerns have made this kind of adventure less possible. Many rockhounds and mid-range dealers also blame the skyrocketing high-end market for dragging up prices at all levels and, from their perspective, ruining the hobby. As we will see in the following section, this complaint is part of a field of conflict between high-end and mid-range dealers and collectors.

Mineral collecting in Mexico was never as popular as it came to be in the United States, though some Mexican collectors, mostly mining engineers and geologists, actively collected in the late nineteenth and early twentieth centuries. Ponciano Aguilar in Guanajuato and José Landero y Cos in Real del Monte were two prominent examples. Small shops similar to U.S. rock shops also sprang up near mining localities and on highways such as the one that runs from Torreón to Ciudad Juárez in northern Mexico. Many of these cater to U.S. collectors or tourists who drive into Mexico to visit mines and to collect: a practice that has tapered off with the increasing politicization of the border and the rise of violence in northern Mexico.

Beginning in the 1970s, Dr. Miguel Romero Sánchez emerged as the preeminent Mexican collector. Indeed, Romero reportedly had a de facto right of first refusal for high-end minerals from Mexico for at least twenty years (see chapter 4). Romero joined a growing company of high-end collectors in the United States and Europe who, along with a group of innovative dealers and curators, have managed to change the ways that minerals are valued. He also worked to promote mineralogy and mineral collecting in Mexico by helping to form the Sociedad Mexicana de Mineralogía, among other things.

However, mineral collecting in Mexico has not grown in the way it has in the United States and Europe. According to the collectors and scientists I interviewed in Mexico City, there are few Mexican collectors and few large-scale dealers. The Museo de Geología sponsored an annual gem, mineral, and fossil show/fair for about ten years up until the late 2000s. About two thousand to three thousand people would attend, both mineral collectors and lapidarists (those who cut and engrave minerals and rocks). The UNAM and the Polytechnic University also have Geology Weeks (Semanas de Geología); these are career fairs that dealers attend to sell specimens and sets. None of these events attract the high-end collectors that Tucson, Denver, Munich, and other shows do.

On an August evening in 2011, I met with a longtime dealer and collector at the pleasingly retro Vips Madero restaurant in Mexico City's Centro Histórico, where waitresses with flouncy aprons and carefully coiffed hair distributed laminated menus with photos of sundaes and enchiladas suizas. My informant, a man of about seventy, worked for many years in geological exploration for the government and for private mining companies, and he also teaches at the UNAM. He has some twenty to thirty clients, including earth scientists, students, lapidarists, and a few serious collectors. He stated that he is one of the few who sells to collectors. In recent years, moreover, he has been forced to reduce his business because of his wife's serious illness and a lack of access to good minerals. The best material, he complained, is controlled by the dealers in the United States, especially Tucson, and is priced too high for most Mexican collectors. Aside from himself, the mineral store Mineralia (also in the Centro Histórico), and one or two other small outfits, he said that "there are hardly any sellers now [in Mexico], and hardly any collectors."

It is hard to know just how many collectors there are in Mexico. There exist few face-to-face or online venues for Mexican collectors to get together to share their interests and information. The dealer whom I interviewed is trying to change that; every other week he meets with four to five people at the Vips Madero to talk about forming a club for mineral collectors, but they have not been able to get very far. He knows, for instance, that there is a collector somewhere in the state of Coahuila, but he has not been able to find him on the Internet. Ruefully, he remarked that it would be easier to get in touch with other Mexican collectors through their contacts in the United States and Europe. The few serious collectors who do exist in Mexico do not have much to do with each other; they are more likely to attend the Tucson and Denver shows and to exchange with foreign collectors than to communicate or trade with other Mexican collectors.

Aesthetic Minerals

Explanations for minerals' high prices and the monopoly of U.S. and European dealers lie partly in a shift in the ways minerals have been valued. Beginning in the 1970s and gaining momentum over the following decades, a new valued quality for minerals began to emerge and stabilize. This new quality—what is called by its adherents *aesthetics*—coalesced at the center of multiple value-making acts as a newly meaningful form of difference. The quality of aesthetics has stabilized to the point where it is now a relatively successful black box. Not everyone agrees that it is the dominant quality for minerals—alternative qualities can and often are imagined and deployed—but claims based on aesthetics no longer need to be made over and over again; they now tend to be readily recognized and understood without a lot of argument or explanation. Furthermore, these claims now act in the world as effective protagonists by raising

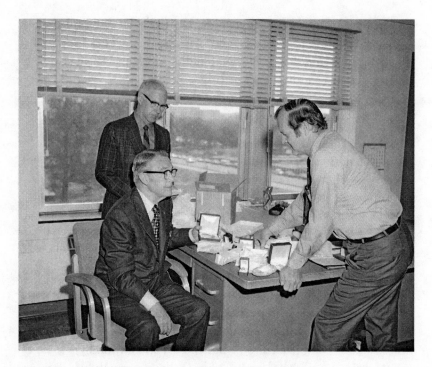

FIGURE 2.3. Paul Desautels (right) with George Switzer and donor.
Smithsonian Institution Archives. Reprinted with permission.

prices and influencing curatorial choices and the actions of miners. Aesthetics has become a relatively durable structure within which minerals are distinguished and ranked. The following section traces some of the specific actors and activities that made this happen.

Paul Desautels: A New Kind of Curator for a New Kind of Mineral

Perhaps the person most directly responsible for this shift in the ways minerals are valued was Paul Desautels (figure 2.3), the curator of the mineralogical collections of the Smithsonian Institution from 1973 to 1983. Desautels came to the Smithsonian from Towson State University, where he taught courses in chemistry and mineralogy. He had already acquired a national reputation as a mineral micromounter (a collector of specimens that require magnification to be properly viewed). In 1957, he was hired at the Smithsonian, and from 1963 to 1983, he was first associate curator and then curator of the Division of Mineralogy, part of the Mineral Sciences

Department.[11] During his early tenure, Desautels did some scientific research and publishing (Desautels and Clarke 1963) but quickly established his strengths in curating, collection building, and fund-raising.

In a report coauthored with Rustam Z. Kothavala concerning the Harvard Mineralogical Museum (see chapter 4), Desautels outlined his view of the contemporary role of mineral museum curating, saying,

> successful operation of a mineral collection is more an administrative function than a scientific one. To be effective today, a mineralogical museum director must be an energetic fund raiser, a skilled operator with people and specimens, a capable manager. In addition to knowing mineralogy, he must know something about public relations and financial management, tax and investment matters, behavioral psychology, public speaking, museology, and the stock market–like activities of the mineral specimen market.

Desautels proved to be remarkably good at most of these functions, and he made connections with the finest dealers and private collectors in the country, as well as with donors on a scale not previously approached by other curators (who tended to be rather staid and unworldly). These capacities helped Desautels build the national mineral collection into one of the most famous and prestigious in the world. In the words of one of my informants who worked with Desautels,

> He was a man of colossal skill, especially with dealing with different kinds of people. Prior to him things were collected and put in drawers and that was fine and next morning the same thing. . . . He was a big collection builder. He had a very discerning taste. He knew that lots of minerals occur in beautiful, beautiful specimens—he could look at a specimen and know it was the best ever found. He was like, "everybody come under the tent." What Desautels did affected the *garimpeiros* ["prospectors" in Portuguese, in reference to the phenomenal tourmalines extracted in Minas Gerais, Brazil] up in the hills. He knew how to say to people "who are you going to leave your collection to—your kids don't want it, but you can benefit your country. You can have your name on the label, you can have your kids' name on the label—you can have anything you want on the label."

Desautels not only responded well to a changing world of mineral collecting; he helped to create that world in several very specific ways. One of the most concrete of his legacies, directly traceable to his influence, is the Tucson Gem and Mineral Society show, which has been running for over fifty years and has become the world's largest mineral event (see chapter 6). In the early 1960s, the early years of the show, its organizers, including Bill and Millie Schupp, invited the Smithsonian to participate and to bring a mineral exhibit. As one informant told me,

> Bill and Millie Schupp didn't know that the Smithsonian didn't travel, and Paul [Desautels] chose to ignore the rule that the Smithsonian didn't travel. He came out and he had a great time. I expect he had a better time socially in the beginning than mineralogically, but he started telling people about it, and they all began to come too.

Desautels's early connection to the Tucson show not only helped to spur the growth of the show itself, but also impelled other mineral museums to enter into the expanding world of mineral collecting in the 1960s and 1970s. As this was also the time when many more people entered the hobby, and when an elite tier of collectors emerged, willing to pay very high prices for minerals, Desautels was able to use these connections to enhance the national collection, while also conferring a certain degree of legitimacy to the hobby.

Desautels also opened up mineral appreciation to a broader public, especially through his books *The Mineral Kingdom* (1968) and *The Gem Kingdom* (1971). *The Mineral Kingdom* was especially influential. Many of the collectors and dealers that I interviewed for this project said that this book had been central to their becoming mineral collectors. Handsomely illustrated by Smithsonian photographer Lee Boltin, *The Mineral Kingdom* became a Book of the Month Club selection and brought appreciation of minerals as striking and beautiful objects to a far broader audience than before. An article in the *Asbury Park Sunday Press* for July 6, 1969, gives some sense of the book's popularity and the ways it drew attention to the beauty of minerals. It describes how a fashion designer, Chester Weinberg, took inspiration for his fall/winter collection from Desautels's book, deciding that "he must have fabrics exploding with giant-sized reproductions of the colorful minerals."

Finally, Desautels cultivated relationships with high-end collectors and dealers and acquired fabulous specimens from private collections. He was adept at bringing in donors as well. His relationship with Marion Stuart (Mrs. E. Hadley Stuart) of Idaho is one example. In the words of one informant, "Desautels knew Marion was a broad who could make it happen (in those days you could say things like that)." Marion Stuart helped to support the Mineral Sciences department and the collections for years, funding research and donating money for specimens. She and her husband E. Hadley Stuart (whose fortune came from the Carnation Milk company) planned to give money for a new gem and mineral hall at the Smithsonian, but they pulled out after Desautels's departure from the Smithsonian in 1983 and donated the money to the Los Angeles County Museum of Natural History.

All of this occurred within a context of rapidly rising prices for some minerals. In Desautels's own words, minerals became "no longer just scientific samples or curiosities. They are expensive and eagerly pursued objects for collectors" (Harvard Report on the State of the Mineralogical Museum, Desautels and Kothavala, 1973, SIA Accession 93–121, Paul E. Desautels Papers, folder 4). Wealthy collectors, enterprising dealers, and a steady outpouring of mineral specimens from old and new localities convened to create new, elite markets for minerals. Minerals became an asset class for investors, and some multimillionaires began to convert part of their fortunes into minerals, further driving up the prices. A good example is the Texas oil magnate Perkins Sams, who—with the help of Desautels—built a spectacular collection and donated it to the Houston Museum of Natural Science in 1985 (Wilson 2012:Paul E. Desautels, 1920–1991).[12]

An article in the *Robb Report* (a "magazine for the luxury lifestyle") states that:

> Thanks to the stock market boom, [the 1990s] was also a period when prices paid for [mineral] rarities skyrocketed. What happened, says Joel Bartsch, curator of the gem and mineral collection at the Houston Museum of Natural Science, is minerals—geological accidents that have miraculously survived the excavation process—came to be viewed as works of art. Where once collectors had chosen specimens based on scientific merit, they now began to select according to aesthetics. (*Robb Report* 2002)

Desautels actively participated in and promoted this newly elite market for minerals through his close relationship with high-end dealers and collectors, the glamour and panache he brought to a formerly rather stodgy pursuit, and, in all likelihood, his adeptness at obtaining high appraisals for mineral donations to the museum, which donors could use for tax write-offs. In all these ways, he helped move minerals from being valued as objects of natural history to being valued as fine art. Desautels's efforts to appeal to a higher "class" of collectors and to emphasize the beauty of minerals were early value-making acts that helped to establish aesthetics as a prized quality.

Preparing the Market: Dealers and the Promotion of Aesthetics as a Value

The *Robb Report* article links the shift to aesthetics as a primary value for minerals to the rising prices, and there is no doubt that likening minerals to fine art meant that people would pay much higher prices for them. Not surprisingly then, a number of dealers began in the 1970s, 1980s, and 1990s to promote this new valued quality vigorously. In doing so, they helped to shift the level of "making difference meaningful" with respect to minerals: that is, they made aesthetic differences between minerals more meaningful than they used to be. One of the earliest of this group of dealers who, like many other dealers, was also a prominent collector was Dave Wilber.

In an interview at the Denver Gem and Mineral Show in 2005, Wilber recounted some of his history as dealer and collector to me. He began as the agent in the West for the New York dealer R. C. Romanella and set up on his own in the late 1950s. Around this time, he decided to spend more money on fewer, more choice minerals, particularly "gemmy" minerals (with large or well-developed, transparent crystals). He was especially fond of tourmalines (a group of crystalline silicate minerals that comprise some of the most expensive collectible minerals in the world). He showed his collection at Tucson and "made a big splash." In 1974, he bought Peter Bancroft's collection for $400,000—the most that had ever been spent on a mineral collection—and in 1982, he sold his own collection, "full of showy, aesthetic, high ticket items" to Perkins Sams for $600,000.

Wilber's claim to fame, along with his spectacular collections and keen business sense, was his insistence on perfection in minerals. Prior to Wilber, several people told me, flaws, chips, bruises, or other signs of damage in minerals were relatively unimportant to the overall value of a piece. But Wilber, in his own words, is "into perfection," and he influenced other dealers and collectors so that they too got into perfection.[13] He was so successful in branding this perspective that a nick in a mineral specimen is now sometimes called a "wilber." For Wilber, however, it is legitimate to restore a damaged mineral to a state of perfection. Here the emerging analogy between minerals as fine art proved especially useful; in describing why repairing a mineral is justified as long as it is professionally and "tastefully" done and honestly reported, Wilber told me, "It's just the same as restoring a painting."

This reference to fine art, particularly painting, is common among dealers (and also collectors). For instance, Stuart Wilensky (2009) writes on his website:

> Minerals are my personal passion. My intention is to share with fellow collectors this passion, the joy we have experienced in owning and handling so many wonderful mineral specimens, and to encouraging the collecting and preservation of mineral specimens as works of art. By displaying these archival photos wherein others can view, compare and enjoy aesthetic mineral specimens, we hope to promote mineral collecting right alongside the more traditional fine arts of painting and sculpture.

Dealers who cater to this end (the high end) of the mineral collecting market have their own shows (such as the Westward Look Show in Tucson and the Dallas Fine Minerals Show), their own publications (such as those produced by the publishing house Lithographie, Inc.), and their own specialized language, such as the phrase "fine minerals," which strategically links minerals to fine art.

Another primary actor in this emergence of aesthetics as a value is the company Collector's Edge Minerals, founded and run by Bryan Lees, first in his class at the Colorado School of Mines and a consummate businessperson. Collector's Edge has made a number of game-changing moves in the mineral market. It is one of the only mineral dealerships that also conducts specimen mining, as it did most successfully at the Sweet Home Mine in Alma, Colorado. The Sweet Home Mine, formerly a silver mine, was reopened in 1991 by Lees for specimen mining. In the 1990s, it yielded the world's best rhodochrosites, including the Alma King, reputed to be the most expensive mineral ever mined in the United States and now on display in the Coors Hall of the Denver Museum of Nature and Science. Collector's Edge has also made mineral collecting history by establishing the first-ever mineral preparation laboratory that specializes in repairing and trimming high-end minerals for collectors (see chapter 5). And finally, Collector's Edge changed the ways that minerals are displayed at mineral shows and in private collections, drawing inspiration from art galleries rather than natural history or science museums. Through flattering lighting, tasteful carpeting, black velvet linings in display cases, unobtrusive Lucite stands and other props, Collector's Edge helped to change the way people looked at minerals. In the process, the company also made a great deal of money.

I have interviewed many of the dealers most closely associated with the shift to aesthetic minerals and it is obvious that they entered the business because they love minerals and feel passionately about them as objects of beauty. There is nothing insincere about these folks' love of beautiful rocks. At the same time, all of them would, I believe, freely admit that promoting the aesthetics of minerals is good business. The opening of a new market for aesthetic minerals has expanded the range of minerals that can be sold, and the analogy to fine art has allowed for much higher prices. Minerals (at least the high-end, showy kind) have become glamorous, and glamour sells.

Preparing the Collector: Connoisseurship

The stability of minerals as aesthetic objects also depends on a public that is willing and able to appreciate their beauty. Collectors, then, have done their part to make the shift to aesthetics as a valued quality by cultivating new capacities of taste and connoisseurship. As in the world of fine art, there is a delicate balance between the presumed subjectivity of responses to beauty and a contentious field within which objects and people struggle to establish themselves as beautiful or as capable of recognizing beauty. As part of this process, a new language of "mineral connoisseurship" emerged in the 1980s, focused on the appropriate capacities of taste and refinement to recognize "exquisite" minerals.

In an influential article published in the *Mineralogical Record* in 1990, "Connoisseurship in Minerals," the journal's editor-in-chief, Wendell Wilson, distinguishes between scientific and aesthetic connoisseurship. Scientific connoisseurship, in his estimation, uses the criterion of exemplarity that goes along with evaluating a mineral specimen; a valuable mineral in this case is one that accurately exemplifies its species, locality, or the laws of chemistry or crystallography more generally. Aesthetic connoisseurship is more complex, and Wilson admits right away that "defining or explaining mineral aesthetics is to some extent an empirical problem: if enough people pronounce something beautiful, then it is beautiful" (Wilson 1990:7). He goes on to identify a number of qualities that have been valued by many, including color, transparency, luster, a pleasing sculptural shape, and so on.

The article is essentially didactic, aimed at aspiring connoisseurs rather than already developed ones. It emphasizes the learned nature of connoisseurship (stressing that "refined" connoisseurship is based on experience) and the degree of mental and sensory labor that it entails. For instance, Wilson describes connoisseurs as "making finely discriminating judgments," "devot[ing] all their personal resources to the enjoyment of aesthetics," and "work[ing] out the hidden mathematics [of a mineral's shape and composition]," phrases that emphasize that mineral connoisseurship is a matter of fine motor cognition, as it were, a painstaking skill in making nice distinctions.

The book *Ikons, Classics and Contemporary Masterpieces,* distributed as a special supplement to subscribers of the *Mineralogical Record,* provides a more recent guide to connoisseurship and its role in "building a world-class collection." Written by Wayne Thompson, one of the most famous and successful collector/dealers, the book first classifies "world-class specimens" into "ikons," "classics," and "contemporary masterpieces" (the latter two, if sufficiently fine, may also be ikons). Ikons are "The ultimate object of comparison. . . . Ikons achieve their lofty status purely on the basis of visual impact, with little or no consideration of provenance, history or individual physical qualities" (Thompson 2007:7). The book also enumerates the necessary qualities for a world-class collector (including knowledge, financial resources, a discriminating mind, and courage). It also describes "desirability factors in mineral specimens": a list headed by "aesthetics" and also including condition, rarity, and—somewhat lower down—provenance (Thompson 2007:11–24).

These descriptions of mineral connoisseurship locate much of the value-making labor with the collector rather than the miner or dealer (Callon et al. 2005; Foster 2007). In fact, as we will see, the presumed pristineness of minerals forms much of their appeal, so it is not surprising that the roles of miner and dealer are minimized (chapter 5 describes the implications of pristineness in detail). It is the serendipity of the minerals' beauty and the capacity of the connoisseur to recognize that beauty that attracts many aesthetic collectors. For example, on his website, the dealer Stuart Wilensky (2009) meditates on the beauty and wonder of minerals, saying:

We can all appreciate the talent and skill of artists like Van Gogh or Michelangelo, but what can we say about the overwhelming beauty and natural aesthetics of minerals? They are truly the wonders of the world. They are the treasures of the Earth, each one is completely unique and cannot be duplicated by man.

The twin ideas of minerals' spontaneous, unauthored beauty and of the need for refined expertise for its recognition help to enlist collectors in the project of making aesthetics a newly important form of meaningful difference.

Views from Outside

Not everyone in the mineral collecting world has gone along happily with this shift toward aesthetic minerals. In fact, many were and are quite disgruntled, even offended, by what they see as the lack of interest in science, the overvaluation of individual ikons or "killer" minerals denuded of their context, and the high prices they see as connected to this shift. One interviewee—a mineralogist—described the trend as "trivial and thrilling and pretty and cute and wow, and I have nothing to do with trivial, cute and pretty," while another described himself, with some pride, as "only interested in nebbishy minerals."

Some collectors complain about the "elitism in mineral collecting" that drives up prices and ignores the scientific and locality information associated with mineral specimens. For instance, one mineral photographer talked to me about a conflict, during the production of a glossy photographic book of a famous aesthetic collector's collection, over how much locality information to include. The collector wanted only general locality names with the photographs, and the photographer wanted to include more specific information about particular mines and pockets.

For their part, those who describe themselves as aesthetic collectors can be very dismissive of their more scientifically oriented cousins. One group of high-end collectors that I met at the 2005 Denver show advised me, laughing, "the three worst things you can do as a collector is collect by species, by locality or in suites [by region]." This statement shows the move, characteristic of the aesthetic minerals movement, away from minerals as exemplary, indexical objects. Many collectors are interested in both aesthetics and science, in varying combinations, but tensions between the two tendencies endure.

The move toward aesthetics has attracted many new people to mineral collecting, raised prices, and promoted the extraction and preservation of intact specimens that are likely to meet the criteria of aesthetic collectors. In this sense, mineral collecting is thriving, and even the grumpiest scientific collector will acknowledge that better specimens have come out in the past twenty-five years than ever before. The rise of aesthetic mineral collecting, then, provides a good example of how the process through which a new form of meaningful difference emerges creates new allegiances, conflicts, protagonists, and forces.

3 MAKING SCIENTIFIC VALUE

In most instances, the relevance of the concept of value to understanding uses of minerals seems obvious. Minerals are often exchanged as commodities and mined as the main objects or the by-products of extractive industries; in these areas, their economic character is immediately evident. To be sure, there are many cases in which minerals are exchanged not as commodities but as gifts either to other humans or to divine beings, but gifts are a favored topic within economic anthropology and are easily seen as objects of value. In other contexts, however, the people involved are not primarily concerned with minerals' movements as gifts or commodities, though they may busily participate in such exchanges. They mostly use minerals to produce or instantiate scientific knowledge of one sort or another. In this context, are questions of value-making still relevant?

Once we specify our understanding of what value actually is—that is, the social action that results in making meaningful difference and in making difference meaningful—the domain of scientific knowledge becomes especially apt for looking at value, because this is exactly what successful scientific practice does. The case of mineral specimens shows this especially well, because minerals straddle those domains that are conventionally related to value in anthropology (commodities, gifts) and those that are not (scientific artifacts, religious offerings).

This chapter uses three extended examples to show how the messy and fitful process of making scientific value has over time, with greater and lesser success, helped to create new kinds of actors such as mineral species, collections, localities, and technologies. These in turn help to create Mexican and U.S. people, places, and institutions, and the relations of power among them.

The cases presented here show two closely linked aspects of value-making. First, value-making can be productively seen as the more or less successful convening of actors into a network that is meaningfully different for other objects around it. A so-called "regime of value" (Appadurai 1986) is not prior but subsequent to that process. Second, this does not mean that value is created new at every moment. Earlier value-making that results in successfully stable convenings of actors (black boxes) form the landscape within which current value-making occurs. This makes some convenings more likely than others. This second part of the argument crucially brings questions of power into play.

My three examples are the discovery of the new mineral species aguilarite; the field exploration and collection of rocks and minerals as part of the U.S.-Mexican Boundary Commission of 1849–1856; and the discovery, documentation, and publicizing of a cave of giant gypsum crystals in the Naica mine in Chihuahua. All three show how value is created through happenings in the production of scientific knowledge concerning Mexican minerals. The three cases vary by time, place, type of scientific or technological activity, and importantly, by degree of success. For if value-making is the more or less successful convening of actors into a network that is meaningfully different for other objects around it, then it makes sense to look at cases where such networks have and also have not been successfully stabilized, as well as cases where we can see attempts at stabilization of difference in process and those where we see formerly stable networks break apart and the actors that make them up disperse.

Aguilarite

In 1890, the Guanajuatense mining engineer and mineralogist Ponciano Aguilar was working as the superintendent of the San Carlos mine in Guanajuato. He had already begun to collect and study minerals and many of the specimens in his collection come from this period.[1] This was just at the beginning of the period when U.S. mining companies began moving into Mexico in great numbers, many of them remaining until the Revolution of 1910 and beyond, to the period of the Cardenista administration of the 1930s. Like many of the mines on the Veta Madre, San Carlos was especially rich in minerals composed of silver, in some cases with other metals, bonded to sulfur.[2] Aguilar discovered a few specimens that appeared to be naumannite (silver selenide) (figures 3.1 and 3.2). In 1890, William Niven, a field collector employed by the mineral dealer George L. English, came to Guanajuato and met Aguilar, who provided

FIGURE 3.1. Aguilarite, Ag₄SeS. Photo by Wendell Wilson. Reprinted with permission.

FIGURE 3.2. Ponciano Aguilar in his study. Archivo Histórico de la Universidad de Guanajuato, Fondo Familia Ponciano Aguilar. Reprinted with permission.

him with a few specimens (Wilson 2006:304). They apparently discussed whether or not the specimens represented a new mineral species (that is, one not yet formally described or named). Upon his return to the United States, Niven gave the specimens to Frederick Genth, a chemist based in Philadelphia. Genth sent it to Frank Penfield, another chemist, with whom he collaborated regularly, and Penfield's analysis showed that in addition to silver and selenium, the specimens also contained sulfur, distinguishing them from both acanthite (Ag_2S) and naumannite (Ag_2Se). On account of this, Genth identified it as "a new species, which has been named Aguilarite in honor of its discoverer" (Genth 1891:401).

On February 14, 1891, Niven wrote to Aguilar, saying:

> Dear friend—I have much pleasure to inform you that the silver selenide has been analyzed by professor Genth and found to be a new mineral, which is named "aguilarite" as recommended by me. It is a most interesting species and I sincerely congratulate you.
> Yours very cordially, Wm. Niven.

Most Mexican minerals (and indeed, most minerals first identified from localities in the Global South) are named for locations or for European or North American scientists or others connected to minerals. The fact that aguilarite bears the name of a Mexican scientist, Ponciano Aguilar, is quite unusual; only four minerals of the sixty-five (as of 2002) first discovered in Mexico are named after scientists of Mexican nationality: aguilarite, ordoñezite, mendozavilite, schmitterite (Gait 1999).[3]

Aguilarite's discovery took place at a moment when U.S. mining companies were just beginning to move into Mexico in a big way. By the 1890s, the United States began moving south to take over Mexican mines. At the time of aguilarite's discovery, the San Carlos mine was under Mexican ownership, but within eight years it and nearly all other mines in Guanajuato were taken over by U.S. companies (Meyer Cosio 1999).

Meanwhile, U.S. museums and universities were consolidating power and influence as centers of knowledge production and stewardship. Public support of the earth sciences in Mexico was returning after a half century of decline (see chapter 2), and Mexican scientists such as Aguilar were again reaching out to international colleagues and institutions. In addition, the interest in the United States in Mexican mineralogy, both for private and university collections and for scientific researchers, was expanding rapidly.

I have begun this story with the people involved, but in fact a whole range of actors converged to make the new object aguilarite, of which only some were human. However, before we go on, we need to understand the process by which a new mineral species comes into being. The Commission on New Minerals and New Mineral Names (CNMNMN), the international committee with the authority to ratify the discovery and baptism of new minerals, defines a mineral species in the following terms:

> A mineral substance is a naturally occurring solid that has been formed by geological processes, either on earth or in extraterrestrial bodies. . . . A mineral species is a mineral substance with well-defined chemical composition and crystallographic properties, and which merits a unique name. (Nickel and Grice 1998)

Even though the CNMNMN was formed in 1959, the definition quoted follows a standard scientific practice that also prevailed in the late nineteenth century, when aguilarite was discovered, and aguilarite, like many (but not all) minerals identified prior to 1959, was "grandfathered" in by the commission. The CNMNMN definition of a mineral species conforms, more or less, to the criteria followed by Genth.[4] For aguilarite to "merit a unique name," it had to be different from its neighbors in meaningful ways, with either a distinct chemical composition or a distinct crystal structure from all other known minerals, or both (Dunn 1977). This meaningful difference needed the coming together of a whole variety of actors, including Niven, Genth, and Aguilar, who came together in their mutual interest in identifying a formerly undescribed mineral and including elements, geological and chemical forces, technologies, and so on.

For instance, three elements convene in the chemical structure of aguilarite: silver, sulfur, and selenium. Aguilarite appears as a middle term in the "acanthite-naumannite series" (Francis et al. 1999), a continuum from Ag_2S to Ag_2Se. The boundaries between these three different species (e.g., acanthite, aguilarite, and naumannite) are not completely distinct; instead, there is a "compositional range." So, some specimens denoted as acanthite contain small amounts of selenium, and some specimens denoted as naumannite contain small amounts of sulfur. Within the compositional range of aguilarite, sulfur and selenium are roughly equal. (According to one study, this range is defined as running from $Ag_4S_{0.95}Se_{1.05}$ to $Ag_4S_{1.10}Se_{0.90}$ [Petruk et al. 1974].) Whereas in many cases the middle term of a binary solid solution series (a series of minerals where one element completely replaces another at the two end points of the series, as with sulfur and selenium here) is not recognized as a separate mineral, the x-ray powder pattern of aguilarite reveals a sufficiently different structure to qualify it as a separate mineral species (Earley 1950:244–245). Over the course of the series, however, selenium completely replaces sulfur. So the existence of the three species and their interaction depends on the substitutability, in chemical terms, of selenium for sulfur. This property of the series helps to bring these species into being and to make the distinction between them meaningful.

Now we can step back to look at the interaction of the different species as part of a stratified geological profile. The substitution of selenium for sulfur in the acanthite-aguilarite-naumannite series in the Guanajuato mining district can be correlated with the sequence of the district's three major geologic formations. In the lowest level, dating from the Mesozoic Era, only acanthite occurs; aguilarite can be found beginning in the second layer of red conglomerate, approximately 1,300 meters thick and dating from the Eocene period (it is known as the Guanajuato Formation); aguilarite continues, accompanied increasingly by naumannite, through a layer of volcanic rock laid on top of the Guanajuato Formation. This paragenesis speaks to the earlier presence of sulfur (prior to 67 million years ago) and the later entrance of selenium (56 million years ago or less) (Francis et al. 1999; Vassallo 1988). The geologic time frame for this part of aguilarite's story is huge, but we can infer a slowly unfolding movement from acanthite to aguilarite and then to naumannite over the course of millions of years.

Once aguilarite was established as a new mineral species, it achieved a new status; it became valuable by virtue of its distinction from other silver minerals present in the Guanajuato district and, more broadly, from all other minerals already discovered, described, and named. Thus a new kind of scientifically significant object—the mineral species aguilarite—emerged at the crux of a new set of associations. It became a black box. Once this happened, aguilarite became an actor in its own right, in the sense that it could now do things. What kind of things did aguilarite do?

First, it engaged Aguilar in new or altered relationships with scientists and collectors. For instance, a year after notifying Aguilar of the new mineral's baptism as aguilarite, Niven wrote again, on February 8, 1892, this time sending a specimen from his own explorations:

I send you herewith a specimen of the new rose garnet which I have
been working since I saw you last. I would like to hear from you. Have
you found any more aguilarite? I would like a nice specimen for my private collection. I will give you more particulars of the prospects of this
country when I hear from you.

Yours sincerely, Guillermo Niven. Wm. Niven

Likewise, Pedro Monroy, one of the premier mining engineers of
Mexico, wrote to Aguilar to "congratulate on the honor of which you
have been the object" and to request a sample (Colección Ponciano
Aguilar, box 98). The fact that aguilarite now existed as a new mineral
made it a desired object for collectors, particularly those who specialized
in minerals from Mexico. The facts that Aguilar was in control of the
only known source of aguilarite and that it had been named for him
engaged him in new transactions and communication both in Mexico
and abroad.

Aguilarite also enlarged the category of Mexican minerals, thus
making Mexico that much more important to the field of descriptive
mineralogy as a whole. To understand why this is, we need to know a bit
more about type localities. In mineralogy, type locality refers to the locality where a new mineral species was first discovered. Once aguilarite
was elevated to the status of a mineral species, distinct from acanthite
and naumannite, it also made the San Carlos mine in the La Luz mining
district into the type locality for that species. As of now, there are approximately sixty-five Mexican species (species for which the first specimen described came from Mexico). Collections that emphasize a particular country or area—such as Mexico—may commit their owners or
curators to collecting an example of as many specimens from that country or area as possible. The most prestigious collections can go one step
further and seek the type specimens (a particular specimen from among
those used to describe the new mineral) from particular localities and
areas.

The existence of aguilarite as a new mineral species also led to new
scientific research. So, a series of studies mapping the presence of selenium at different levels of the Guanajuato mining district, as well as the
identification of a number of other silver minerals with selenium content,
began with the identification of aguilarite in 1891 (Earley 1950; Petruk and
Owens 1974; Vassalo and Reyes-Salas 2007).

Likewise, an article from *Canadian Mineralogist* begins:

> The relationships between acanthite, aguilarite and naumannite were
> investigated by studying acanthite and aguilarite in an ore suite from
> Guanajuato, Mexico, naumannite, aguilarite, and acanthite in a sample
> from Silver City, Idaho, and a synthetic naumannite. (Petruk et al.
> 1974:365)

The communications and exchanges that made those Mexican and U.S.
samples available for study are not especially interesting to the authors
or, probably, to most of the readers. Nevertheless, they had to happen.
Someone had to write, call, or go to Guanajuato, and the samples had to
be brought or sent back. Scientists had to talk to each other and probably also to technicians, secretaries, shipping services, customs officials,
and so on, on both sides of the border. These interactions took place
because the relationships among acanthite, aguilarite, and naumannite
were seen as sufficiently interesting to merit study. If aguilarite had never
been distinguished from acanthite and naumannite, the question of their
relationships would never have been posed, or at least not in that way.

Aguilarite continues to contribute to the production of new actor-
worlds within mineral collecting. Once the correlation between the depth
of mineral deposits in the Guanajuato district and the ratio of selenium
to sulfur was firmly established, new forms of knowledge production
could occur. For instance, in their presentation on silver mineralogy for
the Annual FM-TGMS-MSA (Friends of Mineralogy—Tucson Gem and
Mineral Society—Mineralogical Society of America) Mineralogical Symposium, later published in the collectors' magazine *Mineralogical Record,*
Francis et al. report on microprobe analyses of thirty-one specimens and
confirm that

> the Se/S ratio is indeed diagnostic for determining the proximity to
> the [uppermost] Tertiary volcanic unit. The only specimen of nauman-
> nite investigated was from the uppermost level of the San Carlos mine,
> while the large, lustrous Rayas mine acanthites (with nearly zero sele-
> nium) are from the deepest levels of that mine. (1999:85)

The word *diagnostic* is critical here; the authors go on to recommend
a topographic mapping of the mineralogical distribution in the district,
in part so as to supply more accurate locality data for "the large number
of specimens that have the simple, but unfortunately unspecific, label
'Guanajuato, Mexico'" (p. 85). The ratio of selenium to sulfur, which
came to be known through the process of naming aguilarite and then

differentiating it from its neighbors, could help establish particular specimens through new actor-networks, this time composed of particular specimens, locality data, collectors, collections, and so on.

In this sense, aguilarite as a newly stabilized assemblage of actors exerted pressure in the formation of other assemblages, such as mineral collections, and of Guanajuato as a mineral locality. Telling the story this way shows that aguilarite's birth was not predestined or inevitable, though neither was it completely happenstance or made up out of whole cloth. It was contingent on the coming together of the actors I described (and of others too, such as the stagecoach that brought Niven to Guanajuato and the price of silver that made the San Carlos mine a profitable concern) under particular circumstances. As a result of this successful convening of actors, aguilarite became a black box; the process and rationale of its definition as a distinct species no longer needed to be argued, but instead could be taken as given. It thus became a valuable object and an actor in the stabilization of new assemblages. In fact, the statements "aguilarite has value" and "aguilarite is a stable and meaningfully different assemblage that can animate new convenings of actors" are two ways of saying the same thing (one, admittedly, quite a bit more cumbersome than the other).

This point of view emphasizes the ways in which the social does not exist as an a priori ground on which value-making acts occur but emerges through value-making. However, this does not happen in an empty landscape but rather in one already created and populated by prior value-making acts. Successful value-making results in stable actor-networks, and stable actor-networks help make value-making successful. Stable actor-networks are thus likely to reproduce themselves and to produce the world in their own image and their own interests, resembling in this respect Bourdieu's concept of habitus (Bourdieu 1977).

In the case described here, we can see several ways in which aguilarite's emergence and the aftermath of that emergence solidified inequalities between the Mexican and U.S. actors involved, not initially but over time. For instance, while Aguilar was honored in the name of the new species, this honor seemed to dissolve as the decades passed.

The son of a tailor, Aguilar was born in 1853 in the city of Guanajuato. He received an education in Mining Engineering at the Colegio del Estado de Guanajuato. He graduated in 1876 and went to work for the Negociación Minera la Luz y Anexas. During this first job, he began to collect mineral specimens. In 1879, he married Micaela Zavaleta Pérez-Gálvez, a member of a wealthy and prestigious mining family (Jáuregui 2002:9–15). His marriage brought him into the center of Guanajuato aristocracy, and his local, Catholic education made him an ideal candidate for teaching positions at the colegio and for contracts in mining and public works.[5] In addition to his expertise in mining, Aguilar was a civil engineer who worked on dams, tunnels, railways, and tramways in the city and all over the state of Guanajuato. He also worked periodically as a mining engineer for different companies in Guanajuato, usually as superintendent of mines. In this capacity, he was able to collect mineral specimens and to compose an impressive collection, primarily from Guanajuato. Developing, cataloguing, and otherwise maintaining this collection became some of the central aspects of Aguilar's scientific work.

Throughout his life, Aguilar participated actively in the Mexican and international scientific communities. In 1906, Aguilar helped plan and host a field trip to the mines of Guanajuato for visiting geologists as part of the Congrès Geologique International. In 1912, he attended the first of a series of scientific congresses sponsored by the Sociedad Científica Antonio Alzate. In 1919, he presented his ongoing work on identification of minerals by means of the electric arc (then called "voltaic" arc). In addition to improving his collection, much of his working life was devoted to developing this process and using it to identify specimens in his collection. He considered himself to be a contributor to universal scientific knowledge.

Aguilar maintained diverse international links over the course of his professional life. He received visitors from France, Italy, and the United States, with particular interests in mining, civil engineering, and mineralogy. He was a member of the Rotary Club, the National Geographic Society, La Société Academique de Histoire Internationelle, and the American Association for the Advancement of Science, among other organizations. He often entertained visiting dignitaries, scientists, and travelers when they passed through the city, and he did the same for William Niven.

Niven was a Scottish immigrant to the United States, a self-taught mineralogist and collector of archaeological artifacts. His class position and educational level were far lower than Aguilar's, but the recorded history of the mineral's naming tended to downplay Aguilar's role in favor of Niven's. In this process, the prestige of scientific discovery, like other forms of value, moved from Mexico to the United States. The muting of Aguilar's role as knowledge producer exemplified the intensifying quasi-colonial relations between the United States and Mexico in the late nineteenth century.

Mineral nomenclature follows certain conventions; the procedures of the CNMNMN states that:

> A mineral is commonly named for the geographical locality of its occurrence, for the discoverer of the mineral (although not if he or she is the author), for a person prominent in the field of mineralogy, or for a particular property of the mineral. (Nickel and Grice 1998:11)

Contemporary accounts of aguilarite's discovery gave credit to Aguilar. Genth's original description states that the mineral has been dubbed "aguilarite, in honor of its discoverer." The scientific journal *Nature* reported in 1891 that "a new and very beautiful mineral is described by Mr. F. A. Genth in the May number of *The American Journal of Science*. It was discovered by Sr. Aguilar of the San Carlos mine of Guanajuato, Mexico and has been named after him, aguilarite" (Nature 1891:89).

However, more recent accounts of aguilarite's discovery note that the mineral was named for Aguilar as superintendent of the San Carlos, but they rarely mention that he was a mineral collector and scientist in his own right. The intellectual role he played in the mineral's discovery has been mostly ignored. This fact has not gone unnoticed; one Mexican mineralogist remarked to me that recent accounts ignore Aguilar's scientific research and his active role in collecting the samples of what turned out to be aguilarite and sending them to the United States for analysis. The Webmineral website, which provided the information on naumannite I mentioned previously, has this to say about the origin of the name aguilarite: "Named for Ponciano Aguilar (1853–1935), superintendent of the San Carlos mine, Guanajuato, where the mineral was found." Likewise, a recent article on William Niven remarks that Niven recommended the name aguilarite "in honor of . . . Aguilar, who had been so courteous to him during his visit" (Wilson 2006:304).

In this instance as in others, in the United States the U.S. scientists tend to be seen as the intellectual protagonists in the production of value, and the Mexicans tend to be seen as passive recipients or intermediaries. This was less the case in the early years of the two republics, but as the United States moved into mining and established its own schools of mines, collections, and so on—and as its efforts to control Mexican territory and extract Mexican resources grew—the balance of power shifted in favor of this U.S. view of U.S. scientists as the brains, and Mexican miners, engineers, and dealers as mere conduits for the production of scientific value with minerals. This was clearly not Genth's intention, nor was it taken that way by Aguilar's contemporaries. With the passage of time, Aguilar's status as intellectual author has broken down.

This is due to many factors, but among them are the facts that Niven took the samples to Genth, who published the scientific description in English, in a U.S. journal. The type specimen (the specimen on which the description was based) did not return to Mexico. It was displayed in the World's Columbian Exposition in Chicago in 1893 and then bought by the Field Museum, which was founded after the close of the exposition and which bought many of the exhibits.[6] Aguilar was the authoritative intellectual and material source for aguilarite in his lifetime, but once he died he was relegated to a lesser role. These unequal relations, moreover, affected the kind of science that Mexicans were seen to produce. From being a nexus for the production of universal scientific knowledge, Mexico became a place that was recognized more for producing particular bits of knowledge about Mexico than for its contributions to "universal knowledge." The lives of Aguilar and aguilarite attest to that shift.

The U.S.-Mexican Boundary Survey (1849–1857)

Not all struggles to solidify stable actor-networks work as well as did the ones out of which aguilarite emerged. My second example comes from some decades earlier, at a decisive point in U.S.-Mexican relations and in the comparative positions of the earth sciences in both countries: the end of the Mexican-American War. It explores the different fortunes of the Mexican and U.S. commissions to survey the new boundary after Mexico's defeat and loss of massive amounts of territory.

The Treaty of Guadalupe Hidalgo, signed by Mexico and the United States in 1848 at the end of the Mexican-American War, included Mexico's cession of approximately 1.36 million square kilometers, 55 percent of its territory, to the United States. In exchange, the United States paid Mexico $15,000,000 and agreed to pay any debts of U.S. citizens against the Mexican government. Article V of the treaty described the location of the new boundary in general terms, according to a map published by J. Disturnell in 1847 and included with the treaty, and stipulated that:

In order to designate the boundary line with due precision, upon authoritative maps, and to establish upon the ground land-marks which shall show the limits of both republics, as described in the present article, the two Governments shall each appoint a commissioner and a surveyor, who, before the expiration of one year from the date of the exchange of ratifications of this treaty, shall meet at the port of San Diego, and proceed to run and mark the said boundary in its whole course to the mouth of the Rio Bravo del Norte. They shall keep journals and make out plans of their operations; and the result agreed upon by them shall be deemed a part of this treaty, and shall have the same force as if it were inserted therein. The two Governments will amicably agree regarding what may be necessary to these persons, and also as to their respective escorts, should such be necessary.[7]

This article makes the process of fixing the line sound very straightforward, but in fact carrying out this article of the treaty took eight years and was beset by a nearly absurd number of obstacles, including lack of funds, attacks by Apaches, theft of horses and mules, illness (including the death of the Mexican commissioner, General Pedro García Conde, in 1851), incarceration (for José Salazar, the commissioner appointed by the Mexican government, no less), and intense personal and political wrangling (within the U.S. commission, Congress, and the Department of the Interior) (Hewitt 1990; Rebert 2002; Werne 1987).

The two commissions worked separately, each with personnel and funding from its own government, though they met and collaborated periodically. The U.S. commission was led first by John Russell Bartlett, and then (upon Bartlett's removal when Franklin Pierce came into office) by William Emory. The Mexican commission was led first by Pedro García Conde and later (upon García Conde's death) by José Salazar Ylarregui.

Added to these multiple obstacles was the fact that the Disturnell Map had erred in its placement of El Paso del Norte (now El Paso, Texas),[8] leading to years of confusion and conflict (figure 3.3). The Gadsden Purchase / Venta de la Mesilla (depending on your point of view) of 76,800 square kilometers in 1853, for $10 million, only partly resolved this issue. Other complications—such as the diversion of the Río Grande to a new course and discrepancies in Mexican and U.S. measurements of a "marine league" (Dear 2005)—made the survey both complex and highly contested.

Reams have been written on the political complexities besetting the survey, particularly on the U.S. side, on the topographical work itself, and

FIGURE 3.3. The Disturnell Map, 1847. General Records of the
U.S. Government, RG 11.

on the role of the survey in western expansion and U.S. nation-building.[9]
This more limited account looks at how the different actors on the scene
aimed to create mineralogical and geological value through the scientific
collections obtained on the survey and subsequent publications. If we see
these activities as efforts to create stable actor-networks that could then
become protagonists in further value-making projects, what kind of a
picture do we get?[10] How can we tell this story in a way that brings out
both the contingency in the process of value-making and the relevance
and force of prior value-making acts?

In part, we can do this by comparing the circumstances and relative
success of the two commissions—Mexican and U.S.—called for in the
treaty. Each of these was embedded within its own web of actors (though
not completely independent of the other).

The U.S. Commission

The main tasks of this original boundary survey included surveying, mapping, and marking the line (with makeshift stone markers that
would later be replaced as part of a second boundary survey in the 1890s).

However, Bartlett's official instructions when he embarked on the survey stated that:

> As the organization of the Commission under your charge has been made for the purpose of collecting information relative to the country contiguous to the boundary line . . . it is desirable that you should avail yourself of every opportunity afforded by your passage through the unexplored regions of Texas, New Mexico and California to obtain information as to its geography, national history [*sic*], &c, when it can be obtained without retarding the progress of the boundary survey. (1965 [1854]:590)

Particular notice was made of the utility of mineralogical information:

> The collection of information relative to the precious metals, quicksilver and the various minerals, ores and other substances, useful in the arts, is very desirable. In reference to the existence and localities of these, as well as the locations of mines formerly worked by early settlers in California and New Mexico, and since abandoned, owing to the incursions of the Indians, or other causes, you will do all in your power to acquire information as far as you may be able, without impeding the main objects of the Commission. (quoted in Bartlett 1965 [1854]:590)

Each of these passages, it should be noted, underlined the priority of the "main objects" of surveying, mapping, and marking the boundary line. Furthermore, the special emphasis on mineralogy derived from the potential for economic exploitation rather than any idea of "pure" scientific knowledge.

Given the stated primacy of geological and mineralogical information (botany, zoology, and ethnology were not mentioned in the official orders—except in a general sense if, as I suspect, "national history" is intended to be "natural history,") it is odd that, according to Bartlett's highly entertaining popular account of the survey during his tenure, there were four people assigned to botanical collecting, and three to zoological collecting (and Bartlett himself gathered much ethnological information). Bartlett lamented that:

> as Congress denied me an appropriation for the purpose, I was unable to secure the services of any geologist competent to make such investigations as were desirable. . . . It was highly desirable to institute a continuous series of geological and mineralogical researches, and to form a cabinet illustrative of the structure and mineral resources of the

country; but both the scientific corps and the number of soldiers at my disposal for the purposes of escort, were too small to admit of this being done. (1965 [1854]:ix)[11]

The secretary of the commission, Dr. Webb, one of those responsible for zoological collecting, also collected rocks and minerals and sent some of them back to Washington. In addition, Bartlett reported that "the collection made on the journey from the Copper Mines of New Mexico to California was mostly lost, in consequence of the abandonment of the wagons and the perishing of the animals" (Bartlett 1965 [1854]:550–551). Collections from other parts of the journey did apparently make it home (though these do not appear as separate accessions in the Smithsonian).

The rocks and minerals, along with the more extensive flora and fauna collections, became the basis for a series of federal reports, the most famous of which was William Emory's Report of the United States and Mexican Boundary Commission (1857), which included chapters on zoology, botany, paleontology, and geology. In addition, an artist, Henry Cheever Pratt, made a series of drawings and paintings of scenes from the boundary region (The Albuquerque Museum 1996).

The Smithsonian Institution received most of the collections from the U.S. Boundary Commission and incorporated them into a larger plan to create a national museum. In part, this was due to the quiet but dogged efforts of Assistant Secretary Spencer F. Baird to amass natural history collections (largely to increase pressure on the institution's secretary, G. Brown Goode, who did not favor a museum). The many exploring expeditions sent out from the 1850s to the 1880s to map the North American continent sent back railroad cars of new plants, insects, birds, and rocks. The Smithsonian's hoard grew by over ten thousand accessions a year, and by 1863, some 86,847 collections had been entered in the catalog, each containing many specimens (Henson 2005). The collections of the U.S.-Mexican boundary survey were among these early acquisitions.

There remains little record of the impressions of the collectors on the survey or of their correspondence with Baird, to whom they were sending specimens. A few letters from John H. Clark reported collecting some birds, fish, and other zoological specimens, along with lively complaints about the country ("of all the barren, waterless regions on the face of the earth, I want to see no more than I experienced on this route") and the commission itself ("I never saw such indecision and want of purpose in any undertaking as is manifest in this commission"). Little mention is made in particular of rock and mineral specimens, though these do show up in accession records at the museum, along with plants, animals, and fossils (SIA Record Unit 7002: letters from John H. Clark to Spencer Fullerton Baird, July 4, 1851, and November 13, 1851).

The survey did manage to make a name (literally) for a number of species of plants and animals and to serve as the basis for further scientific research on these. The Smithsonian Annual Report for 1856 (the year the survey was finally completed) touted the botanical and zoological finds, saying (with a dose of hyperbole),

> The Mexican Boundary Survey has imperishably identified itself with the history of the progress of science in collecting perhaps a larger number of new species of North American animals and plants than any one party ever gathered before, or will again. (Smithsonian Institution 1856:43)

The first volume of Emory's report includes a general geographical and geological description of the region covered by the Boundary Survey and a report by James Hall based on collections made by C. C. Parry. Hall prefaces his report with a disclaimer as to the scientific contribution of the report, saying "Our knowledge of the geological formations of the west is not so rapidly progressing, and the materials accumulating in such abundance, that whatever may be presented today as new and in advance of previous knowledge, will to-morrow be regarded only as a historical record of our progress" (Hall 1987 [1857]:105).

He follows this introduction with a list and description of specimens, arranged by segment of the Boundary Commission's route. On the basis of these specimens, in comparison with specimens from other regions, Hall suggested that the area of the boundary survey east of the Sierra Nevada consisted of metamorphic rocks of a later age than those in the East and West of the continent, which exhibited similar lithological features. Hall was also able to surmise several oceanic inundations in the West, succeeded by volcanic activity. Finally, he examined some of the specimens from mining areas of New Mexico and Sonora, suggesting the likelihood of mineral resources in nearby regions not yet explored (and these predictions were borne out spectacularly by the silver and copper bonanzas in Arizona in the 1870s and 1880s). As was often the case with the earth sciences (more than with botany and zoology), Hall's report moved back and forth between pure scientific description and discussions of the economic possibilities reflected in the geological and mineralogical record.

Although the report describes specific specimens with detailed locality information attached, these collections did not survive intact. It is impossible to identify which specimens pertained to which accession, and, in most cases, the rocks and minerals were included in accession with plants, animals, and ethnological artifacts.

Here the multiple actors involved included Smithsonian administrators, curators, registrars, and scientists; the accession records themselves; various men on the commission; the secretary of the interior who issued the collections; the Apache who stole mules and otherwise threatened the commission; the microbes that sickened the commission members; the congresspeople who failed to make appropriations for a geologist, Dr. Webb; the animals who died and wagons that were abandoned; and the rocks and minerals themselves, whose weight likely impeded their easy travel back east. The lack of some actors also made a difference; in the 1890s, the railroads and stagecoaches that led from Guanajuato north probably helped in the discovery of aguilarite, but no such conveyances existed in the border region in the 1850s (indeed one of the purposes of the survey was to find a possible southern transcontinental route for the railway).

The geological and mineralogical collections gathered on the Boundary Survey resulted in several publications as part of Emory's report. These seem to have had little lasting effect as actors in the world. Google Scholar's "cited by" function (which shows which later works have cited the Emory report) yields thirty-seven citations, covering fish, yucca, marriage, mollusks, and Spanish heritage, but nothing related to geology or mineralogy. This need not have been so; one could imagine a situation in which the reports and the collections on which they were based did become more identifiably distinct (and therefore potentially more influential) actor-networks. Had the aims of Baird, Bartlett, Webb, and others been fulfilled and a "cabinet illustrative of the structure and mineral resources of the country" (Bartlett 1965 [1854]:ix) been successfully constituted, with its own accession numbers and perhaps a distinctive identity as the "U.S.-Mexican Boundary Survey Collections," these collections might then have formed the basis of new mineralogical descriptions and other acts of scientific value creation.

If this had happened, it would have been based on the successful distinguishing of some rocks and minerals from others, as took place with aguilarite. Not only did this not take place, but in most cases, rocks and minerals were not clearly distinguished from other natural history objects collected on the survey but were dumped into the same accession with plants, animals, and other items. Even these accessions were not clearly distinguished from collections gathered as part of other governmental surveys at the time, such as the explorations of the northern route for the railroad and of California (both also mentioned in the Smithsonian report for 1856 [43]). Because they were not constituted as meaningfully different from other collections, they could not act as potently as they might otherwise have done.

The Mexican Commission

As hiccupy and problematic as the process of creating value with the rocks and minerals collected by the U.S. commission was, it fared better than efforts on the Mexican side—these barely got out of the gate.

This was not for lack of trying. García Conde, upon his appointment as commissioner, recommended that a mineralogist, zoologist, and botanist be attached to the survey, with funds from his alma mater, the Colegio de Minería. His request was not granted, but the surveyors and topographers assumed the roles of naturalists. José Rosal, who was appointed engineer, was to take care of botanical and zoological collections but soon resigned the position, and José Salazar Ylarregui, also a graduate of the Colegio de Minería, was to act as geographer and mineralogist (Hewitt 1990:177). However, to my knowledge, no trace of collections made by these men and no publication concerning the natural history of the border based on data from the boundary survey exists.

Once José Salazar became commissioner, he renewed the request for dedicated, paid naturalists, asking for one botanist/zoologist and another geologist/mineralogist as well as a medical doctor, but he was told, "it is not necessary for the principal object [of the commission] and should be omitted, considering the current state of finances" (Salazar Ylarregui n.d.).

In addition, the commission suffered from grave deficiencies in other areas, including faulty equipment and the lack of a sufficient military escort. Because of these gaps, at certain points in the survey, notably in the survey of the Rio Grande portion of the line, the Mexican commission was forced to rely upon the U.S. commission's calculations, or to follow behind the U.S. commission and approve their calculations in advance. At one point, Emory asked the U.S. government to pay some of the Mexican commission's expenses, saying, "The Mexican Commissioner himself is eminently qualified to perform his duties, but he seems to be left with means wholly insufficient" (quoted in Werne 1987:208). In 1852, an article appeared in the Chihuahua newspaper *El Chiste* reporting that the commissioner had arrived in Chihuahua City to ask for funds from the government to alleviate "the penury in which the Commission finds itself, on account of its abandonment by the government" (Ramírez 1890:360–361).

Political upheaval in Mexico also contributed to problems for the Mexican commission, most acutely when Salazar was thrown into jail by the Liberal government that ousted General Santa Anna in the Revolution of Ayutla (Ramírez 1890:208)[12] in 1855 (in fact the presidency changed hands four times in that year), throwing the surveying of the line and any related scientific endeavors into disarray.

Furthermore, because the survey's historical circumstances meant that its significance for the U.S. team was diametrically opposed to its significance for the Mexican team, enthusiasm for gathering knowledge about the border region also differed. While the U.S. team was exploring new territory in hopes of finding resources and building infrastructure, the Mexican team was inscribing its nation's defeat onto the landscape. Tamayo and Moncada write, "We can conclude that no more disheartening situation exists in the country [Mexico] than recognizing the loss of half the national territory and having to conduct the work that will formalize that loss" (Tamayo P. de Ham and Moncada Maya 2001:96; translation mine).

In prior years, the Mexican government had had the resources and will to invest in natural history exploration. From 1826 to 1829, the Mexican government conducted a survey of the border region (farther north than the area traversed in the 1850s) and employed a French naturalist, Jean-Louis Berlandier, to make observations and collections in natural history. Ironically, that collection was purchased by the Smithsonian Institution in 1853.[13] No doubt, however, collecting information predicated on the loss rather than gain of territory was less appealing. At any rate, the multiple actors involved in the Mexican survey did not come together in such a way as to produce mineralogical knowledge or new actors of value (such as species, reports, collections, or technologies). The lack of these emergent actors made it difficult for the contributions of the Mexican team, even in areas considered more central to the work of the commission, to be recognized. This was true both at the time and later, and it shows how the success or failure of attempts to make value can make possible or foreclose future value-making projects.

Even in the areas considered central to the work of the commission, the contribution of the Mexican engineers has until recently been consistently underplayed. In his report, William Emory stated that:

> in this operation I looked for little or no aid from the Mexican commission, for although composed of well-educated and scientific men, their instruments were radically defective. Our determinations, after being re-observed and re-computed by the Mexican commission, were received by them without correction. (1987 [1857]:5)

Even as the two teams were engaged in a joint project, the labor of one side (the United States) was seen as more scientifically valuable than that of the other. Though this was at least partly because of material circumstances—the Mexican team was smaller, with less and poorer equipment and far less funding than the U.S. team had—it had lasting repercussions on the kind of science the Mexican team was able to produce.

As in the case of Ponciano Aguilar in the discovery and naming of aguilarite, the role of the Mexicans in the production of scientific value was reduced from author to amanuensis. In large part this was because the reports of Bartlett and Emory became the definitive accounts of the commission. The journalistic and scientific reports in these books, both of which were published in the 1850s and made widely available, not only helped to set contemporary opinions about the survey, but also provided sources for historians and scientists. Salazar and Francisco Jiménez also wrote reports and diaries of the survey, but neither of these was published and only a few copies still exist. Reports, documents, and archives formed under unequal conditions of production contribute unequally to the formation (or lack thereof) of new actor-networks. As Michel-Rolph Trouillot has put it, "inequalities experienced by the actors lead to uneven historical power in the inscription of traces" (1994:48).

Indeed, we can see this in the new research conducted on the Mexican commission following the discovery of new documents in the Archivo Histórico Genaro Estrada (Hewitt 1990). The emergence of these new records—and a political shift within the discipline of history—have led to a reevaluation of the role of the Mexican scientists, recognizing that the Mexican commission conducted considerable independent work, without which the boundary survey could not have been completed (Hewitt 1990; Rebert 2004; Tamayo P. de Ham and Moncada Maya 2001; Werne 1987). The newly available documents, now recognizably distinct not only from other documents that are not focused on the survey, but also from the U.S. documents, have re-energized efforts to establish and recognize the scientific work of the Mexican team as valuable.

The Crystal Caves at Naica

My final example for this chapter concerns a series of caves found in the mines at Naica, Chihuahua. These caves, which contain extraordinarily large specimens of gypsum (of the variety called selenite), have helped to produce not only scientific knowledge, but also new technologies. In addition, where the other two examples focused on the circulation of specimens and collections (as well as publications and other objects), the Naica caves have also famously produced narrative descriptions, photographs, PowerPoint presentations, and films that have become protagonists in the formation of actor-networks. In these multiple reproductions of the caves in different media, actors struggle to lay claim to knowledge not only for themselves but for their institutions, their states, and their nations. Because this case is the most contemporary one, we can see these struggles unfolding. The case shows current attempts to counteract relations of inequality between the United States and Mexico that earlier uses of the caves reproduced and helped to cultivate.

Silver was discovered in the Naica hills in southeastern Chihuahua at the end of the eighteenth century, but a mining claim was not issued until 1896. In 1900, mining operations began, producing lead, zinc, and silver. The Naica mines were closed from 1922 to 1935 because of the revolution and subsequent unrest. The deposit was mined on a small scale until 1961, when Peñoles reopened its operations there. The mine complex grew to be the largest lead producer in Mexico, though production has been falling steadily over the past five years. In January and February of 2008, workers went out on strike, finally gaining a 6 percent wage increase (Reuters 2008).

In 1910, miners in the Maravillas mine at Naica discovered a large cave full of enormous crystals of selenite (the colorless, transparent form of crystallized gypsum), with individual crystals measuring up to six feet long; the huge cave came to be called La Cueva de las Espadas (the Cave of Swords). The cave was located 120 meters below the surface (Foshag 1927; García-Ruiz et al. 2007). In 1912, a description of the cave was published by N. Degoutin in the Mexican journal of the Sociedad Científica Antonio Alzate (Degoutin 1912).

The cave was maintained by the Peñoles company, with limited access to visitors and some attempt to discourage theft and vandalism. It received little international attention until 1927, when a joint Harvard-Smithsonian party of scientists visited Naica as part of a mineralogical fieldwork and collecting expedition. On August 6, 1927, Harry Berman (the member of the team from Harvard) wrote to his boss, Dr. Charles Palache (professor of mineralogy and curator of the Harvard collections), saying:

> We [Berman and William Foshag of the Smithsonian] have received permission from the Peñoles company to remove two large groups of gypsum crystals here at Naica. The first group we chose is the finest by a long way of all the groups in the cave. It is the most magnificent mineral group I have ever seen. . . . We've had four men working three days to get it to the surface. . . . The group consists of fourteen crystals of five and a half feet or more, all terminated and of a beautiful pearly luster. The group is known as the 'maguey plant' and it lives up to its name. . . . The other group is not really of the same class, but it ranks next in the cave, I think. The men start work on cutting it up on Monday. The hiring of the men and building of boxes and packing material will probably cost us about a hundred and fifty dollars. (Harvard University Geological and Mineralogical Museum archives, letter from Harry Berman to Charles Palache, August 6, 1927)

William Foshag, curator of mineralogy for the Smithsonian (see chapter 2), published a description of the cave in *American Mineralogist* in 1927. He described the cave in enthusiastic terms:

> [The gypsum crystals] grow from the floor of the cave in a manner resembling the maguey plants so common on many of the hills of Mexico. These large crystals are colored a light gray by included mud but are often capped by a clear white termination. They are somewhat rounded on their prism zones but are bright and shiny. The walls are studded with scattered crystals of selenite of a stumpy habit. Further on, the way leads thru a narrow opening, just large enough to permit the passage of a man, completely lined with blade-like crystals and forming a veritable corridor of swords. . . . From this passage one descends a few feet into the largest chamber of the cave. The floor ahead rises at an angle of about 30° and is completely banked with myriads of selenite blades one to three feet in length. At the crest of the floor there arises a remarkably fine radiated group of crystals over four feet high, gray in color but tipped with white and glistening brightly in the light of the lamps. (Foshag 1927:254)

The Naica selenites are remarkable not only for their huge size, but also for the opportunity to see them in situ, and it is this that Foshag seeks to capture in this rather uncharacteristically detailed and effusive description. However, between 1927 and 2000, the cave attracted little attention and did not generate new research or publication, though the specimen groups in the Smithsonian and at Harvard were seen by millions of people and certainly contributed to the value of those collections (see chapter 4, figure 4.2).

In 2000, two brothers, Juan and Pedro Sánchez, discovered another cave with gypsum crystals while drilling for a new tunnel for Peñoles. This cave, at 290 meters below the surface, is considerably farther down and far more spectacular than the Cave of Swords. It is known as the Cave of Crystals and has a temperature of 130° F and a humidity index of 100 percent; the prism zones of the crystals in this cave have grown to 36 feet in some places. Adjacent caves, now known as the Queen's Eye Cave and Cave of the Sails, were also found at this time. Since 2000, the caves have been visited by a number of scientists, documentarians, and others engaged in explaining the phenomenon of these giant crystal forms and preserving and publicizing the caves (figure 3.4). Industrias Peñoles signed a formal agreement with the Mexican nongovernmental organization Speleoresearch and Films (S+F) to help preserve and document the caves. In 2006, Peñoles also formed an agreement with La Venta, a group of Italian speleologists that is collaborating with S+F.

The Naica crystals' enormous size makes them meaningfully different from other gypsum formations (which typically do not grow beyond a meter). This distinction results from the interaction of a whole range of actors, including elements, compounds, heat, humidity, and molecular structures. The Spanish geologist Juan Manuel García-Ruiz and several coauthors wrote an article for *Geology*, attributing the crystals' spectacular growth to a "self-feeding mechanism driven by solution-mediated anhydrite-gypsum phase transition at a temperature slightly below the gypsum-anhydrite equilibrium temperature" (García-Ruiz et al. 2007:329). Both anhydrite and gypsum are made primarily of calcium, sulfur, and oxygen in the form $CaSO_4$. In the case of gypsum, there are two water molecules bound to each $CaSO_4$ molecule. The transition between the two minerals occurs at 136° F. Below 136° F (58° C), the mineral anhydrite takes on water and becomes gypsum ($CaSO_4 \cdot 2H_2O$), whereas above 136° F it remains stable as anhydrite ($CaSO_4$). García-Ruiz

FIGURE 3.4. Scientists taking DNA samples from gypsum crystals, Naica, Chihuahua. Photo by Giovanni Badino / La Venta Esplorazioni Geografiche. Reprinted with permission.

and his colleagues argue that the temperature at this level of the Naica mine must have been stable at just below 136° F for many hundreds of thousands of years, allowing for the continuous growth of the gypsum crystals. In the Cave of Swords, closer to the surface, the temperature fell sooner and crystal growth ceased (p. 328). Thus calcium, sulfur, oxygen, and hydrogen interacted under very specific circumstances in such a way as to produce the gigantic gypsum crystals that made the caves famous.

The extreme environment of the Cave of Crystals (where the largest gypsum formations can be found) means that people can only remain inside for a few minutes without protective gear. Indeed, before Peñoles installed security doors at the entrance to the cave, one worker tried to enter (presumably to extract crystals) and was discovered some hours later steamed to death. María Antonieta Ferreira, the director of the Naica project for S+F, remarked in an interview on the risks associated with entering the cave, "if you faint, you're not getting out of there alive, because people won't be able to carry you out in time."

The interaction of high heat and humidity caused the crystals to grow so spectacularly, and in turn, motivates the invention of new technologies, because if the crystals were not so amazingly large, no one would need or want to subject themselves to the dangerously extreme environment. S+F and La Venta have been working to develop safety equipment and instruments to allow people to stay longer in the caves so as to research the selenite formations properly. They have designed cooling suits and jackets lined with ice to protect the vital organs from heat (interview with María Antonieta Ferreira, October 2010), as well as safety protocol and camera equipment adapted to the heat and humidity (National Geographic Channel 2010). The suits have allowed researchers to remain in the cave for up to an hour (without protection, humans can only remain in the cave for ten minutes, because the near 100-percent humidity prevents the body from sweating).

One researcher reported after a trip into the caves on May 18, 2007:

> Despite the cooling breathing systems and the Ptolemy suits, we could not stay there for more than half an hour, but we still managed to collect the rock samples and place them into the drum. As for the physiological aspects, recovery time has been quite long, as the accumulated heat had started to "cook" us slowly, without us even realizing it. (La Venta 2007)

By later that month, teams were able to spend up to an hour in the caves, with the help of a fast freezer unit that allowed them to repeatedly cool their protective suits. Thus, the selenite crystals and the desire to conduct research on them as a unique phenomenon spurred the creation of new methods to keep bodies from overheating.

Once objects are created as valuable, they often go on to create further copies or representations of themselves, in the form of verbal descriptions, photographs, films, paintings, and so on. The Naica caves have produced a lot of these. Narrative and visual reproductions of the cave and the crystals have circulated widely. These serve to broaden the influence of the caves as actors in the world. Narrative descriptions published on blogs and in magazines (Hogan 2002; London 2003) and visual representations of the cave, such as PowerPoint slide presentations (for example, Naica Crystal Cave n.d.) and videos posted on YouTube and elsewhere, have furthered the caves' virtual circulation.

Publicizing the caves is one of the main objectives of S+F, which has produced a film on the caves that was released in theaters in Mexico and on the National Geographic Channel in December 2010. S+F sees the wide circulation of images of the caves as their best protection; this organization aims to get the caves declared as national patrimony and hopes that someday an on-site museum can be installed so that the caves could be viewed safely, through glass. This could be done, S+F points out, without disrupting mining production. However, the company "does not value [the caves]," Ferreira told me. "What they're interested in is mining silver." Ferreira sees the production of films, slide shows, and photo exhibits as the best means to attract resources for the caves' protection. She has appealed to the government of the state of Chihuahua for help, pointing out that the caves could provide good publicity for the state at a time when the world associates northern Mexico only with drugs and violence, but so far Chihuahua has not offered financial or other forms of support. She said, "When you call attention [to the caves] you raise their value," so that now that the film is released, perhaps the company and government will contribute resources to the project and take action to protect the caves.

The caves were also the subject of a photographic exposition mounted on the fence surrounding Chapultepec Park in Mexico City (this space is maintained by the city as a public art gallery). The exposition, which ran in July and August of 2010, consisted of 120 photographs of high production quality by the Mexican photographer Oscar Necoechea, documenting the caves, the research team and its technical equipment, the miners, and the town of Naica. The magazine *Milenio* (2010) reported that Elena Cepeda, director of the Instituto de la Cultura, "expressed her hope that the next governor of Chihuahua might have the sensitivity to value this patrimony of humanity and create an on-site museum at Naica." Like S+F, she aimed at using the visual reproduction of the caves to increase their value in the world and thus to attract interest and resources from other actors.

Cepeda and S+F, in particular, also hope to increase the caves' value as one of Mexico's natural wonders. By displaying the photos at the "Rejas de Chapultepec," a open-air gallery in the heart of Mexico City, and by arranging theatrical releases of the film in Mexican theaters, the Instituto de la Cultura and S+F aim to reclaim the caves from a status of "world patrimony," housed in the United States, to that of Mexican patrimony located on site at the mine and in the capital city. This creative maneuvering of a "language of patrimony" is common in many parts of the world but especially strong in Mexico (Ferry 2005b), and, in this case, it has the potential to intervene as an effective actor or force in its own right. Here, however, the caves are not showcased as a "natural wonder of Mexico" to be interpreted and managed by scientists and institutions in the United States or Europe (as we saw in the treatment of the earlier caves), but in equal partnership with Mexican scientists and Mexican institutions. For instance, the photographer, whose name is clearly displayed as auteur, is Mexican; Mexican scientists are displayed in the photos alongside Italian, Spanish, and other scientists; and Mexican institutions join foreign institutions in the credits at the end of the show.

As we have seen, Mexican authority in the earth sciences has always been recognized within Mexico, but, since the mid-nineteenth century, it has often been obscured or erased in the United States. To the extent that this shifting of claims over the caves from the Smithsonian and Harvard museums and experts to Mexican institutions and experts succeeds, it can also open up new possibilities for making Mexico a place of scientific research and global stewardship. That is, value-making occurs in a landscape already shaped by earlier value-making, but it can also help to shift or shape that landscape for the future.

These cases show the multiple interactions of entities of different kinds, themselves in the process of convergence and dispersal. Successful convergence resulted in emergence of an object or piece of knowledge as meaningfully different from other comparable objects or pieces of knowledge, and therefore as valuable. One type of actor that has played a role in making value in this way is the collection, in which individual minerals are temporarily integrated or subsumed. Collections, however, are a particular kind of corporate entity for which processes of value formation work in a particular way. The following chapter turns to these processes.

4 MINERAL COLLECTIONS AND THEIR MINERALS: BUILDING UP U.S.-MEXICAN TRANSNATIONAL SPACES

Mineral collections—intentionally assembled groups of minerals held out of economic exchange and manifesting exemplarity, aesthetics, or some other value—are sites where interactions and transactions between Mexico and the United States are especially densely clustered. Furthermore, the intentional character of collections brings a self-conscious quality to these interactions and transactions. This is particularly true of "Mexican mineral collections," that is, those that purposefully focus on minerals originally found in Mexico, either freestanding or as parts of larger collections. Collections are expressions in miniature of particular visions of the United States and Mexico that are brought into being through the interactions and transactions of the people and objects that make it up.

Viewing collections in these terms is potentially fruitful for two reasons. First, the objects in these collections are chunks of Mexico, and extracting them in a form that can be sent to other places necessarily involves mining, territorial claims, and other engagements with Mexico as *material* space. Second, the mines and mining corporations, scientific institutions, marketplaces, and collectors' networks within which minerals from Mexico circulate are mostly populated by those from Mexico and the United States and operate mostly within Mexican and U.S. *social* space. Collections of Mexican minerals act as a meeting place for these material-social relations, these things and people, and a stage for the production of U.S.-Mexican transnational space. The capacity of collections to express U.S.-Mexican space in this way emerges both intentionally, through the explicit and implicit claims to value made by collectors and curators, and unintentionally, as the tidemarks of economic, political, and institutional arrangements that bring minerals into collections.

Most scholarship on U.S.-Mexican transnational space has focused on the activities of recent, often relatively poor migrants, and their everyday domestic, working, and affective practices. It does not typically focus on transactions between Mexicans and Anglos from other social strata or from other areas of life (such as scientific research or collecting). Many of these works have great force and insight, and the transnational lived space they describe reflects many people's experience. However, viewed as a whole, this body of work tends to prioritize lived space built up over time through the activities of working-class and peasant migrants, imaging it as *the* definitive U.S.-Mexican transnational space. This discussion aims at a more plural understanding, and the three collections described here can be seen as the ground on which three different versions of transnational space have emerged. However, to see how and why collections might be seen this way, we need to know about what collections are and how they work.

"The Collection Is the Thing"

Collections—groups of objects intentionally gathered together, arranged, and displayed according to some nonutilitarian logic or logics—emerged in Early Modern Europe and were closely tied to the rise of science as a set of techniques for apprehending the world (Daston and Park 1998; Findlen 1995; Pomian 1996) and to imperial encounters (Mitchell 1991; Thomas 1994). In the emergence of the two North American nations that form the subject of this book, both scientific and imperial expansion became a central concern of elites. It is therefore not surprising to find lively traditions of mineral collecting in both Mexico and the United States.

A primary effect of collections is the assertion of value. How do they assert the value or values that they do? They do so by imposing new forms of order and understanding on disparate things, drawing together objects from diverse contexts into a system of classification and in doing so creating a new context that takes precedence over the objects' origins, though they may draw on those origins as one of their criteria of inclusion. So, for instance, a collection of ethnographic objects may select objects on the basis of their originally intended uses (as weapons or clothing, for instance), but upon entrance into the collection, these objects are no longer meant to be wielded or worn. The defining values for mineral collections have shifted over time, from curiosity to exemplarity to pristineness.

Collections also create value by themselves becoming valuable objects, corporate beings made up of, but transcending, their constituent parts. In part this happens through the function of collections as enclosures that keep out the time and space of the world outside the collection (Baudrillard 1994; Stewart 1993). To illustrate this, let me give an example.

A major component of the U.S. national mineral collections (part of the Smithsonian Institution) is the collection of Washington A. Roebling, donated to the museum in 1926 by his son, John Roebling.[1] Along with the minerals themselves, John Roebling established a fund for purchases and field collecting. These new specimens become a part of the Roebling Collection in the National Collections. I asked the director of the department of Mineral Sciences and head curator of the national collections whether it was possible to distinguish between the original Roebling collection specimens and more recent acquisitions. He responded:

> Well, you can always go back to the records, and even by looking at the accession number—the original Roebling specimens cover a certain range [of accession numbers], so you know if it's outside that range it's a more recent acquisition. But in a sense we don't want to make that distinction. The argument that it's not a dead collection, but it's acquiring more and more pieces every year that are labeled as Roebling specimens is a powerful one. *The collection is the thing*—and it's continuing to grow.

This discussion of the Roebling collection as "not a dead collection," but one that can grow over time with acquisition of new specimens, resonates strongly with many collectors' understanding of what makes collections compelling. Collections in this view are living, dynamic entities that exist above and beyond individual pieces and can confer value onto those pieces.

Assemblages or actor-networks are corporate entities that take on a life of their own. What is the difference between a collection and an assemblage? The first part of the answer is that collections are intentionally brought together, but so are things like electric vehicles, universities, and mineral species. The intentionality of collections has a special character. The things brought together are claimed as similar in some way, and, second, that similarity is taken as the purpose of their convening. They are not brought together to do something else, but simply to be in the same place asserting their similarities (and their differences). In other groupings of objects, which I am calling *assemblages,* either the parts are seen as heterogeneous and/or the logic of similarity (the criteria on which the assemblage is based) is not taken as an objective in itself. Thus faculty in a university or mechanics in an auto shop may all be intentionally brought together on the basis of their skills and experience; by my definition these are not collections because the similarities are not meant to be convened for their own sake, but intended (it is supposed) to produce something, such as scholarly research or well-repaired cars. This feature of intentional similarity-for-its-own-sake makes collections a good site for looking at how value-making works, and with what effects, because people involved in making collections are so explicitly and purposefully participating in the process of making meaningful difference and deciding which differences are meaningful. Collections show value-making in a distilled form.

In this chapter, I focus on three Mexican mineral collections in Mexico and the United States, examining the ways in which, in the course of making value, these collections also produce miniature versions of U.S.-Mexican transnational space. I explore both how these three collections create value within their own borders, through the process of including some minerals and excluding others, and how they project their value beyond their own borders in competition with other collections. The first process shows the traces of interactions and transactions concerning minerals and other actors in the United States and Mexico. The second shows collections themselves as actors in cooperation and confrontation with other collections. The two processes (the entrances and departures of particular pieces in a collection and the valuing of collections in relation to each other) together help to build up particular iterations of U.S.-Mexican transnational space.

Mineral Collections and Minerals: Collectivity and Inequality in Transnational Space

Although it is obviously true that many relations of inequality pertain between Mexico and the United States and that these are expressed in some form in almost every domain of experience and activity, depicting U.S.-Mexican relations as solely defined by inequality and exploitation tells only part of the story. The three collections I discuss here show how relations of inequality between Anglos and Mexicans operate alongside more equal exchanges of knowledge and production of value between the two nations, among people engaged in what they see as commonality of purpose transcending national boundaries or occupying a status as transnational elites who can move unproblematically between Mexico and the United States. The circumstances under which minerals enter and leave collections show the range of these relations of collectivity and

FIGURE 4.1. Museum of Mineralogy, Guanajuato School of Mines. Photo by Elizabeth Ferry.

inequality. By exploring the stories of how these three collections came to be and how minerals entered into and departed from them, we can see this range in its full complexity and irreducibility.

La Colección Ponciano Aguilar: Transnational Collectivity of Science

My first example is the personal collection of Ponciano Aguilar (the discoverer of aguilarite). As we saw in the previous chapter, Aguilar was a prominent figure in Guanajuato's elite, a gentleman scientist who dedicated himself to research, mineral collecting, and the cultivation of national and international connections to other scientists. He worked on building his private mineral collection over most of his life, and it served both as a source for scientific research and as a representative collection of Guanajuato, and secondarily Mexican, minerals (figure 4.1). Because Aguilar worked in mines and lived in a mineral locality, many of his specimens come directly from the mine. From the beginning of his career, in his capacities as mining superintendent and civil engineer, he was able to collect mineral specimens and appears to have bought many of his minerals from miners or others (as far as I know, there were no dealers working in Guanajuato until the 1940s). Many of the specimens entered in his notebooks have a price recorded next to them, usually between 50 centavos and 2 pesos, giving evidence that he bought the specimens from others instead of or in addition to collecting them himself. The notebooks do not have a date attached to them, but if we put them somewhere in the beginning of the twentieth century, we can see that these are handsome prices for miners, most of whom made no more than 1 peso per day.

Aguilar's mineral collection went hand in hand with his scientific work. For instance, he used the collection to develop a new method of identifying minerals by means of electric arc (then called "voltaic" arc [*arco voltáico*]). Much of his working life was devoted to developing this process and using it to identify specimens in his collection. Among his personal papers appears a letter from the Universidad Iberoamericana in Mexico City, dated June 27, 1975. It appears to be responding to an inquiry by Aguilar's daughter concerning this work on mineral identification:

> In the technique of pyroelectric analysis using the voltaic arc for mineral identification invented by your dear father, the beginnings of the current method of atomic absorption spectrophotometry [AAS] that has invaded laboratories and have perfected techniques for the quantitative and qualitative analysis of minerals can be discerned. The fact that in 1930 your noble father had made a substantial change in the methods of [mineral] analysis demonstrates the mind of an inventor who moves ahead of his time, so that men in later years might follow the path he set in developing improved techniques. (Colección Ingeniero Ponciano Aguilar Frías [CPA], box 98)[2]

In addition to representing the minerals in the Guanajuato district and in Mexico more broadly, the collection served as a working resource for his scientific and technological work, helping Aguilar in his contributions to a universal science.

Aguilar was a person with deep international ties. Though he never lived outside of Guanajuato, Mexico, he corresponded with many scientists and collectors in the United States and Europe. His files are full of correspondence with mineralogists and scientific societies, and mineralogical articles and citations in Spanish, English, and French. As a member of Guanajuato's elite, his tastes were also international. The invitation to a dinner held in his honor on the fiftieth anniversary of his graduation was printed entirely in French and included such items as *dindoneaux truffées* (turkey stuffed with truffles; a dish that one would be hard-pressed to find in Guanajuato today). This is consistent with the Francophilia of upper-class Mexico in the late nineteenth and early twentieth centuries.

Although his taste in food may have leaned toward France, the transactions that built up Aguilar's mineral collection were more centered in the Americas, and most of his correspondence and exchanges were within Mexico and in the United States. The collection and his personal papers bear the traces of a dense network of contacts between mineral collections in both countries; they are the material form of a transnational collectivity of collectors and scientists.

As his collection grew, Aguilar also purchased specimens from dealers, such as the Comptoir Mineralogique et Géologique from whom he purchased 200 specimens, mostly European (CPA, box 64). And, like most collectors, Aguilar also built his collection through exchanges. The status of his collection and its reputed strength in Mexican minerals and silver minerals, in particular, helped establish these networks. As new minerals crossed the boundary into Aguilar's collection, his networks with scientists, dealers, and collectors became more dense and his collection's reputation more widespread. This in turn led to further acquisitions. For instance, Lazard Cahn, a prominent U.S. mineral dealer, sent a form letter on July 26, 1907, no doubt sent to many mineral collectors and dealers in Mexico and elsewhere, to request a number of Mexican minerals, including aguilarite, and to offer specimens in exchange (CPA, box 56). In 1924, M. M. Vonsen, a grain and feed dealer from Texas, wrote to invite Aguilar to exchange specimens (CPA, box 44). Such transactions were (and continue to be) an important means by which collectors diversify and improve their collections. The collection's reputation brought Aguilar further into Mexican scientific circles. In 1903, the director of the Instituto Geológico de México, Ezequiel Ordóñez, wrote to ask for examples of building stones and materials from the state of Guanajuato, cut into 1-decimeter square blocks. Later that year, another functionary of the institute wrote to remind Aguilar, and also to ask for a specimen of aguilarite for the institute's collection (CPA, box 59).

Aguilar's reputation as an astute collector and a man of science allowed him to cultivate these contacts and to use them to further his

FIGURE 4.2. Minerals Exhibit Hall, Harvard Museum of Natural History (with gypsum group from Naica, Chihuahua on left. © President & Fellows, Harvard College, www.hmnh.harvard.edu. Reprinted with permission.

collection and his research, which in turn helped him to cultivate more contacts. Minerals acquired through exchange, gift, and purchase from Mexico and the United States constitute the material residue of the national and transnational networks within which he moved. When he died, he left behind him a mineral collection containing over nine thousand specimens that was especially strong in silver minerals and calcites from Guanajuato but with many other Mexican minerals as well.

The Harvard University Mineralogical Collections: Traces of Empire

The mineral collections at Harvard rank among the most famous and prestigious in the entire world (figure 4.2). They contain multiple smaller collections within them, and they serve multiple purposes. The world-class excellence of the Harvard collections, however, was achieved relatively recently; the study of mineralogy and mineral collecting had a slow and fitful start at Harvard, as in other places in the eighteenth- and early nineteenth-century United States (Greene and Burke 1974), and the collection did not really become great until the early twentieth century.

Although Harvard had been in possession of a small natural history collection since at least 1750 (on display at Harvard Hall), a separate mineral collection was not identified until the time of Benjamin Waterhouse, who began teaching geology and mineralogy at the Medical School in 1788 and who established a separate collection of rocks and minerals. In 1793, Harvard received its first substantial mineral collection from John Coakley Lettsom of London (Frondel 1988:9).

Waterhouse left Harvard in 1812, probably on account of political and personal differences with his colleagues and the administration; Clifford Frondel, Harvard professor of mineralogy and the collections' curator from 1946 to 1977, described Waterhouse as an "irascible and contentious person, jealous and ambitious, never on good personal terms with his colleagues in the Medical School and the College, and an ardent supporter of Jefferson in a stronghold of Federalists" (Frondel 1988:6). At this time, John Gorham picked up instruction in geology and mineralogy. There was at least some effort to build the mineral collection by this time, for Gorham tried to acquire the valuable mineral collection of General George Gibbs for Harvard, but the collection eventually went to Yale (Greene and Burke 1978:72). In 1824, Gorham was succeeded by John White Webster, who taught until his conviction and execution for murder in 1850. Webster was convicted of murdering and dismembering George Parkman, to whom he owed money, an event that inspired at least one dramatic account and a PBS film (Schama 1991; *American Experience: Murder at Harvard* 2009). Webster's collection played a central role in this crime, for Parkman apparently confronted Webster upon learning that the latter had offered his private mineral collection, already leveraged as collateral against a loan from Parkman, for sale (Frondel 1988:17).

In 1847, instruction in practical mining and geology began at Harvard and continued sporadically, with the Gold Rush of 1849 and the mining bonanzas in Newburyport, Massachusetts, in 1874–1878 and Maine in 1876–1881, providing periodic impetus. The establishment of mines in the West after the Civil War, the foundation of the U.S. Geological Survey in 1879, and the building of the transcontinental railroad all fomented interest in mining, which in turn led to its inclusion in the curriculum of elite eastern universities such as Harvard (Columbia University founded the first school of mining and mineralogy in the United States in 1864). Furthermore, by the end of the nineteenth century, Boston had become the center for mining financing, which improved attention to mineralogy, mineral collections, and mineral exploration (Frondel 1988:8).

As U.S. mining companies began to expand westward, mining elites from the West began to send their sons to Harvard, and Harvard graduates began to make their fortunes in either the exploitation or the financing of mines. These people tended to value metals and minerals, and they also tended to bequeath what they valued to Harvard. Most notably, in 1913, Albert F. Holden (Harvard 1888), a mining engineer and administrator and owner of one of the largest copper companies in the Americas, bequeathed his mineral collection to his alma mater.

According to his obituary in *Mineralogical Magazine,* Holden began collecting minerals in 1895 and continued until his death. His collection contained around 6,000 specimens and included a number of smaller collections (*Mineralogical Magazine* 1914:117–118). He also left a sum of $500,000 for specimen purchases. This bequest helped to establish the Harvard collections as a significant U.S. mineral collection. In turn, as with all successful claims to significance, this helped to attract further bequests and to impel further acquisitions. More concretely, the funds for purchase established the means to make further acquisitions.

Within the larger category of the mineralogical collections, the status of the Mexican minerals at Harvard also grew. In 1892, the curators purchased the first Mexican mineral in the collections, a Mexican opal, from George F. Kunz. The Holden bequest included a number of Mexican minerals. Among these was a collection purchased from Severo Navia, Ponciano Aguilar's teacher and predecessor at the Guanajuato School of Mines, which included 161 specimens and focused on silver minerals and calcites from Guanajuato (Metropolis 1999).

In 1927, William Foshag from the United States National Museum (of the Smithsonian Institution) and a young mineralogist from Harvard, Harry Berman, traveled to Mexico to collect specimens. They visited mining localities all over northern and central Mexico and returned with two groups of gypsum specimens from the Cave of Swords at Naica, Chihuahua (Foshag 1927:23). As we saw in the previous chapter, one of these groups entered the Harvard collections, further enhancing the value of the collection in the mineral collecting world.

In 1947, Alfred Burrage, a mining engineer and industrialist, donated his mineral collection to the museum; the Burrage collection included thirty-eight gold specimens from Mexico and a suite of calcites from the

La Luz district in Guanajuato. In the 1960s, Earle Collins purchased Mexican agates on behalf of the museum, and in 1994, Harvard and the University of Arizona jointly purchased the collection of Christopher Tredwell consisting of 365 Guanajuato silver minerals and associates (Metropolis 1999). Tredwell was working in the nearby city of Irapuato for an affiliate of Ford Motor Company in the 1990s, and used the opportunity to visit Guanajuato frequently. The Santa Fe Cooperative, which controlled many of the city's mines at that time, allowed him to enter several of the mines, especially the Valenciana, for the purpose of collecting. Later, Tredwell was forced to sell his collection for $22,000. Dr. Carl Francis, curator of the Harvard Mineralogical Museum, was interested in the collection because of the already existing strength in Guanajuato minerals through the Navia and Burrage acquisitions. Thus the presence of particular specimens and clusters of specimens within the collection helped to attract further purchases and gifts in related areas.

For instance, I decided to donate an acanthite and silver specimen I had been given in Guanajuato in 1998. The fact that Harvard had a globally recognized collection and a minor specialization in Guanajuato motivated, in part, my donation.

When Dr. Francis saw the specimen, he said, "I've seen specimens like this from Kongsberg, but never from Guanajuato. I like it so much that I'd like to have it for the collections." After I had given the specimen to Harvard, I received a letter of thanks from Dr. Francis, who wrote, after explaining its mineralogical significance:

> The specimen is also important in terms of our holdings. The catalogue lists six Guanajuato silver specimens. . . . Three, like yours, are from specifically named mines but are only modestly interesting. Yours leaps to the head of the list!

Note the mutually constitutive relationship between the collections' value and the specimen's value. First, Dr. Francis emphasizes his liking for the specimen, saying, "I like it so much that I'd like to have it for the collections" (that is, it is worthy of being included in the Harvard collections—something I was delighted to hear). In the letter, the conferral of value goes the other way, because the specimen's intrinsic interest (very slightly) improves the Guanajuato collections.

The traces left in the Harvard collections of transnational movements between the United States and Mexico differ from those in the Aguilar collection. In the Harvard collections, through the movements of Harvard alumni and donors (such as Holden and Burrage) and scientists (such as Berman), we see the rise of U.S. influence in Mexican mining in the end of the nineteenth century and the growth of U.S. mineralogical and earth science research in the early twentieth century. The collections show the traces of expanding U.S. imperial interventions in Mexico in which Harvard men were prominently featured, often carrying off rich rewards from their activities south of the border.

In these traces, with the exception of Severo Navia, relatively little record of the contributions made by Mexicans in mines, museums, or laboratories remains in view, though such traces were surely there. Even my own contribution continues a tradition of moving minerals from Mexico to the United States, with more credit given to those from the United States who helped to move them. Though the deed of gift does mention my friend who gave me the specimen, the original miner who extracted it is not reported (for I was never told who it was). My husband and I appear as the primary human protagonists. In this way, a particular relationship between the two nations—rooted in empire and based on the extraction of value and the erasure of Mexican authoritative action—is built up over time.

The Miguel Romero Collection of Mexican Minerals: Elite Transnational Space

My final example is the collection of Dr. Miguel Romero Sánchez, which ranks as the most famous and valuable ever built by a Mexican collector (figure 4.3). Dr. Romero was born in 1925 in Tocalá, Oaxaca, where his family owned a farm and sugar mill. He attended the UNAM, earning a degree in chemistry, and worked in the laboratory of Eduardo Schmitter, one of Mexico's most accomplished mineralogists of the twentieth century (see chapter 2). After getting further degrees at Harvard and Imperial College in London, Romero returned to private business in Mexico. He eventually joined his brothers and sisters in the poultry business. Grupo Romero (now Grupo Idisa, directed by Romero's youngest son, Alejandro) is an agroindustrial corporation dedicated to developing feeds, medicines, and other products for livestock. Dr. Romero's background in chemistry helped the business to become hugely successful. Dr. Romero became a prominent member of society within the city of Tehuacán, the state of Puebla, and nationally, where he served as member of Parliament (Wallace 2008:6–7).

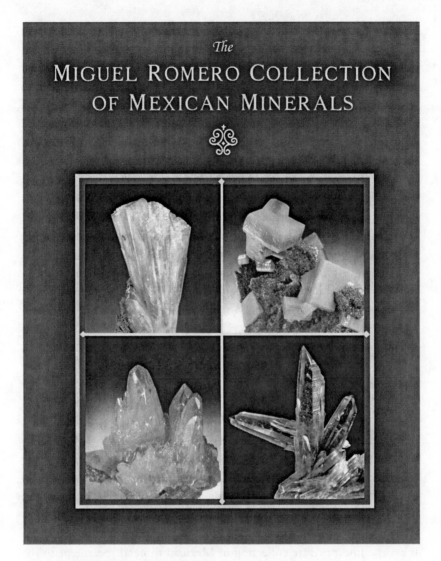

FIGURE 4.3. *The Miguel Romero Collection of Mexican Minerals,* special publication of the *Mineralogical Record. Mineralogical Record /* Wendell Wilson. Reprinted with permission.

Romero began collecting in the 1960s and focused his collection on the minerals of Mexico in the 1970s. According to one of his Mexican colleagues, he began to focus on Mexico after seeing a mineral in the Smithsonian with the locality description "Tetela de Ocampo," with no state name included. Furthermore, there was no record of this mineral in

Mexico. Because there are three towns by that name in Mexico, Romero set out to discover from which of the three the mineral came; it turned out to have come from the state of Puebla. He "saw that more was known in other places [than in Mexico] about Mexican minerals. He decided to dedicate himself to rescue the mineralogical heritage [*acervo*] of Mexico."

During these years, Romero began to attend the Tucson Gem and Mineral shows, where "he was delighted to find a huge variety of Mexican minerals for sale" (Wallace 2008:9). These were the years when Mexican minerals began to emerge in large quantities, usually passing through Tucson by way of El Paso. Romero and a few other collectors and dealers with particular interest in Mexican minerals helped to foment this trade.

In addition to purchasing minerals at Tucson and from the increasing network of specialized, high-end mineral dealers, Romero also traveled to different mining localities to collect in the field, and exchanged specimens with other dealers and collectors. He discovered two new minerals: *mapimite* and *ojuelaite,* for which the type locality is the Ojuela mine in Mapimí, Durango. His collection contained some of the finest specimens to have emerged from Mapimí, including the legrandite specimen known in the United States as the Aztec Sun and several adamite specimens (see chapter 6).

By the end of Romero's life, his collection contained nearly 8,000 specimens, including around 6,500 from Mexico, including 80 specimens of native silver, 90 Naica fluorites, and 120 Mapimí adamites. The collection contained both a systematic series (a collection aimed at representing the different classes of minerals in a systematic fashion) and a display series, including some of the most famous Mexican specimens in the world. The collection thus brought together a comprehensive series of Mexican minerals and also spectacular individual pieces. Terry Wallace describes it as "the finest private collection of Mexican mineral specimens in the world" (Wallace 2008:5).

Romero was a true transnational elite. Educated at Harvard and fluent in English, he was close friends with many of the high-end collectors who emerged in the 1970s, 1980s, and 1990s, as well as with many curators, dealers, and others in the mineral world. Many of the most famous and expensive Mexican mineral specimens ever found have passed through his hands. He came from a comfortable background and amassed a fortune through his skill as scientist and businessman, and he converted that fortune, in part, into Mexican minerals.

Described by his Anglo and Mexican friends and colleagues as an "Old World gentleman," and "a gentleman in all senses of the word [*un caballero en todos los sentidos de la palabra*]," he was clearly recognized as a Mexican collector, but in a very different way from the miners and small-scale dealers in places like Guanajuato or Tucson, not to mention the common stereotypes of Mexicans that circulate in places like Tucson (and, to be fair, all over the United States). His own and his family's experience of U.S.-Mexican transnational space is fundamentally different from that of many migrants or relatives of migrants. This elite, cosmopolitan version of transnational space was both depicted in and in part created through the mineral collection, which brought prestige to Romero and helped him establish contacts at the very top of the mineral world.

Mineral Collections and the World: Universality and Patrimony

Collections strive also to distinguish themselves from other collections outside their boundaries. The three collections discussed in this chapter have each struggled to establish themselves in a transnational field of value, with varying degrees of success. Sometimes mineral collections claim value on the basis of *universality,* the quality of contributing to a universal project such as scientific research, or of representing the mineral world in a comprehensive way. At other times, collections can invoke a "language of patrimony" (Ferry 2005b) in some way, claiming either that the collection represents the inalienable possession of some collectivity, such as a city, nation, or institution (the mineralogical patrimony of Mexico, for instance), or that the collection *itself* is an inalienable possession, a treasure that should be passed down intact to future generations. These two qualities—universality and patrimony—may overlap or they may be incompatible. The first two collections discussed here demonstrate different configurations of the two qualities, whereas the third seeks another solution to the problem of value-making. It seeks neither universality nor patrimony, but finds value in its accommodation with market exchange.

La Colección Ponciano Aguilar: Aiming for the Universal, Settling for Patrimony

Through the course of his life, at least as far as can be told from his personal papers, Ponciano Aguilar gradually amassed a respected collection through a relatively smooth process of value accretion, thereby securing his own position as a respected scientist and collector, both in

Mexico and to some extent in the United States as well. After his death, however, recognition of the collection's scientific (and economic) value became more contentious.

When Aguilar died in 1935, he left four unmarried daughters without financial support. The collection was fully catalogued, first by Aguilar and then by his daughter María Aguilar, with details on provenance for most of the specimens. Over the next several years, José Aguilar, Ponciano's son, attempted to sell the collection in the United States to help provide for his sisters. He engaged an agent in the United States, John Glenn of the Panamerican Trust Company, to advertise the collection to universities and other potential buyers and to broker the sale, if possible. Glenn sent letters to various commercial and scientific institutions in the United States and entered into preliminary negotiations with several parties.

The most extended negotiation was with Ward's Natural Science Establishment in Rochester, New York, the largest supplier of scientific collections and supplies for educational uses in the United States. On April 14, 1938, J. L. Vance wrote, saying, "We are prepared to acquire a collection like the one you describe, and if you could give us complete information concerning it, we will be pleased to see if it meets our needs."[3] Vance asked for specific information concerning the specimens and their provenance and suggested that perhaps Aguilar could send a catalogue to Rochester for their review. Later however, upon seeing the catalogue and hearing the proposed price, a representative from Ward's wrote:

> We think that perhaps you made an error in giving the price of $22,000 for the collection, because that price is excessive. We cannot consider buying the collection at such an elevated price. We are interested in knowing more about the collection, and for this reason if one of our representatives has to visit Mexico, he will come to examine it. However, the collection will not be considered by us at the price mentioned in your letter. (CPA, box 95)

Glenn asked Aguilar if he would lower the price, and Aguilar responded that he would sell it for as low as $2,000 (which gives some indication of the family's financial need), but that Ward's would need to send a representative to see the collection first, adding, "once the collection has been seen and examined, I am completely convinced that only in this way will they be able to judge it at its true value" (CPA, box 95). The correspondence between Glenn and Ward's ends at this point, and because the collection is still in Guanajuato, the proposed transaction apparently went no further.

A number of other parties expressed some interest in the collection around this time, including several dealers and the American Museum of Natural History. One dealer, Martin Ehrmann, noted that he had heard about the collection from several people, including Herbert Whitlock, the curator of mineralogy at the American Museum, and Peter Zodac, a collector and the founder of the magazine *Rocks and Minerals.* However, in spite of the buzz, these discussions also seem to have petered out.

One speculates, though it is impossible to prove, that some of the U.S. collectors and curators may have found it hard to believe a collection worth over $20,000 could possibly come from Mexico. Although there is no contradiction between a collection's universality and its Mexican provenance, some of those on the U.S. side seem to have had trouble accommodating the two, just as some had trouble seeing Aguilar as the intellectual discoverer of aguilarite.

On January 25, 1942, the newspaper *El Universal* reported that the state government of Guanajuato had purchased the collection in exchange for a pension for life for Aguilar's daughters. The article reported that "this collection . . . added to the collection already at the Colegio del Estado, will be one of the best in the country, according to experts" (CPA, box 35).[4]

We can see negotiations between the Aguilar family and various U.S. people and institutions as efforts to establish the scientific value of the collection, as well as the price as the monetary representation of that value. That process was unsuccessful in the United States, at least to the degree originally hoped for by the Aguilars, but it was more successful within Guanajuato.

The collection was not, at that point, able to build its fame beyond Mexico.

As the 1942 newspaper article indicates, the acquisition of the Aguilar collection filled out the mineralogy museum of the Guanajuato School of Mines, which included the collections of the mineralogists Vicente Fernández and Severo Navia (those that are not now in the Harvard collections), as well as a European collection donated by Alexander von Humboldt around the 1830s and another purchased during the gubernatorial administration of Florencio Antillón in the 1870s. The museum (with a mineral collection of 22,450 specimens) is still at the Guanajuato

School of Mines. It now bears the name of longtime professor and museum curator Eduardo Villaseñor Söhle. For years, it received little public attention, especially from outside the city.

Recently there have been renewed efforts to build appreciation for the collection as "mineralogical patrimony" (interview with Mónica Morales Zarate, July 2007; Torres et al. 2006). Such efforts place emphasis on the collection itself as an entity with significance for a given collective (in this case either the city of Guanajuato or the Mexican nation). For instance, an article in the Guanajuato paper *El Correo* (2009) is headlined "Museo de Mineralogía, de los mejores de AL" (Mineralogy Museum, one of the best in Latin America). Efforts to spread the fame of the museum included a published monograph including history and photographs of the collection (Morales Zárate et al. 2008), as well as several articles aimed at informing scholars in the earth sciences of the value of the collections (interview with María Guadalupe Villaseñor Cabral, July 14, 2010).

The monograph is copublished by the Canadian mining corporation Goldcorp Mexico, the Mexican Society of Crystallography, and the University of Guanajuato. The photographer, Jorge Sandoval, follows the style made popular among mineral collectors in the United States of photographing the minerals as if suspended in air (see chapter 5). The black, white, and silver design of the book echoes the dominant colors of the calcite and silver minerals for which Guanajuato is famous. The book opens with a "Letter of Introduction" written by the director of Goldcorp Mexico, which traces the history of the company, including its recent merger with Glamis Gold, "combining the talents of the two best gold mining companies of recent years" (Morales Zárate et al. 2008:1). The prologue ends by exhorting, "This book reflects the commitment of our company to promote culture and education. We hope you enjoy it and we invite you to learn more about the fascinating world of Mineralogy" (Morales Zárate et al. 2008:2). In this way, the company attempts to link itself to the value of the collections as representatives of "culture and education," thus underscoring the collection's value as historical artifact and didactic tool.

This exhortation aims at a broader use and scope for the collection, beyond its status as patrimony. However, the book has reached only a few people, and indeed I was only lucky enough to get it when one of the authors gave me a copy. The transnational aspirations of the Aguilar collection were frustrated, at least as a collection of universal scientific value. As Mexican or Guanajuato patrimony, it has been lately somewhat more successful.

The Harvard Collections: The Universal as Patrimony

The Harvard collection, by contrast, has achieved transnational, even planetary success, both as a contributor to universal science and as the site of spectacular single specimens, such as the gold specimens that are among the world's most highly prized. The collection's great fame arises in part from the fact that it has provided the ground for much scientific research. In 1973, a report commissioned by Harvard on the state of the mineral collection stated that "of approximately 2,000 different mineral species discovered the world over since ancient times,[5] over 5 per cent were first described from the Harvard collection by Harvard investigators— mostly in the past 60 years" (SIA Accession 93–121, folder 4). The fact that the collection had helped to generate so many new mineral species (and therefore retained the type specimens for those species) further enhances its prestige in the scientific world. Providing the ground for producing value, it becomes that much more valuable itself.

This report was written by Paul Desautels (the Smithsonian curator of minerals) and Rustam Z. Kothavala on the occasion of Clifford Frondel's retirement. The report's purpose was to evaluate the collection's "status in the field of mineralogy and what should be done with it" (SIA Accession 93–121, folder 4). The report described the collection thus:

> By historical circumstance Harvard is the custodian of one of the four greatest mineral collections of the world [the others being the Smithsonian, American Museum of Natural History, and the British Museum of Natural History]. If Harvard decides to neglect its mineral collections it would mean that the University has elected to sacrifice one of the world's important cultural, educational and scientific treasures.
>
> It is doubtful if any single collection has contributed more to the development of the science of mineralogy. In a literal sense, it is an irreplaceable national heritage containing many of the finest specimens known, many of them unique. It contains hundreds of specimens on which research has been conducted leading to professional publication and to the development of mineralogical theory. (SIA Accession 93–121, folder 4)

This description sums up the status of the collections both as a museum and as a resource for scientific research, ratifying their place at the very highest tier of world mineral collections.

131

The Desautels and Kothavala report suggests that the Harvard collections achieved both universality and patrimonial status. Phrases such as "one of the world's important cultural, educational and scientific treasures" and "irreplaceable national heritage" stress the cultural and scientific dimensions of the collections, distinguished by the inability to express these forms of value in terms of price and the obligation to preserve the collection for future generations (SIA Accession 93–121, folder 4). The report invokes a language of patrimony emphasizing the collection's status as inalienable possession with an accompanying moral imperative to stewardship (Ferry 2005b). The report uses this language to good effect in the sense that the authors succeeded in persuading Harvard to hold on to its collection and provide for its appropriate maintenance and curation.[6] And indeed, because of that decision, the Harvard collection is still considered at the pinnacle of world mineral collections.[7]

The Romero Collection: Market Accommodations

The Romero collection achieved fame in a different way from either the Aguilar collection or the Mexican minerals in the Harvard collections. Though Romero was an accomplished researcher in mineralogy and the discoverer of two new species, his collection was not notably a foundation for scientific work. While he did put together an impressive systematic collection (one arranged according to standard chemical classification), in keeping with the values of the aesthetic minerals movement he tended to focus on amazingly beautiful and striking individual specimens. Dr. Romero's collection did not invoke the kind of universal value that was achieved by the Harvard collection or aspired to by the Aguilar collection. Neither in terms of its comprehensiveness nor its contribution to science did it claim universality.

Nor did the collection's value, for the most part, reside in its status as Mexican patrimony. Romero did donate part of it to the city of Tehuacán, Puebla, where it is housed in a cultural center. A DVD produced about this part of the collection in 2000 emphasizes its educational value and its representation of the mineralogical patrimony of the country. However, this is a relatively small and unimportant part of his collection. The best and most striking pieces, those that were considered by most of his contemporaries as the heart of the collection, were housed at the University of Arizona and sold by his heirs after his death.

Romero apparently did not intend to keep the collection within Mexico. Instead, he decided to donate his collection to the University of Arizona Mineral Museum in Tucson, in part because of its proximity to the border and in part because Tucson attracts the entire mineral collecting world to its showcase in January and February. However, Romero died in 1997 without having completed the legalities of handing over the collection.

After some ten years of consideration, during which time the display pieces remained at the University of Arizona's museum, Romero's widow and three children sold the most distinctive (and expensive) part of the collection, with the help of Rob Lavinsky of the mineral specimen dealership called The Arkenstone. Little information on the prices these pieces fetched is available, but, for instance, the Aztec Sun sold to a private collector for, reputedly, $1.7 million. The Mexican locality suites remained at the University of Arizona, whereas the Arizona specimens were bought by a prominent Arizona collector and 5,500 pieces from the systematic collection were donated to the UNAM, Romero's alma mater. The Romero collection as such no longer exists.

Though the collection did not primarily claim value as Mexican patrimony, it was recognized as distinctively Mexican, in several respects. For one thing, Romero explicitly specialized in "Mexican minerals" (those found in Mexico). Immediately after his death in January 1997, several members of the TGMS established a special trophy in his name for the best Mexican specimen at the TGMS show. Terry Wallace, one of the people who proposed the trophy, writes "we wanted a good excuse to make sure we examined every Mexican mineral on exhibit at the Tucson Show, just as Miguel would have done" (Wallace 2008:10). The existence of this award both emphasizes Romero's status in the Tucson mineral collecting world and suggests the particular role of Mexican minerals at the Tucson shows. Mexico and Arizona are the only geographic areas specified in the TGMS awards. Furthermore, several of his friends and colleagues report that Romero started collecting because he thought knowledge of Mexican mineralogy within Mexico, compared with the United States, was severely lacking. The collection was intended to address that lack.

The collection was also recognized as Mexican because Romero himself was Mexican. In the commemorative issue on the collection published by the *Mineralogical Record*, Wendell Wilson writes,

It was a tremendous advantage for Miguel Romero to be virtually the only high-end buyer of specimens in Mexico. He was known to everyone, and many fine specimens unearthed in recent decades were brought to him first. Furthermore, even many American [U.S.] dealers who obtained fine Mexican minerals would often give him first refusal, in large part because they respected what he was trying to do in building a national collection within Mexico (Wilson 2008:11).[8]

The value of the collection resided, most would say, in the amazingly beautiful and singular Mexican minerals Romero collected from the 1960s to the 1990s. In successfully claiming value on these grounds, the collection achieved something different from either the Aguilar or the Harvard collections (at least from the perspective of some in the United States): transnational scope and Mexican-ness, in terms of both the character of the collection and the authorship of Romero as collector. The Aguilar collection aimed at universality but was "put in its place" by some U.S. curators, collectors, and dealers such as Ward's Natural Science Establishment, possibly because they perceived universality and Mexican-ness as incompatible. The Harvard collections achieved universality, but mostly at the expense of the Mexican miners, scientists, and collectors who helped bring Mexican minerals into the collection. The Romero collection is recognized by collectors and dealers on all sides as both world-class and Mexican.

However, along with the recognition of the collection as cosmopolitan went the possibility of its dissolution through sale. The sale of the Romero collection disappointed many members of the U.S. collecting community, particularly those who felt that it went contrary to Dr. Romero's stated wishes. Some felt that the family had sacrificed the value of the collection in favor of the monetary reward from its most expensive pieces. Those who felt this way, we could say, see the Romero collection as a classically anthropological "sphere of exchange" that can accommodate morally neutral conveyances (movement of minerals in and out of the collection, in exchange for other minerals) but not morally tainted conversions (liquidation of specimens into cash) (Bohannan 1955).[9]

The breakup of the collection motivated a special supplement to the *Mineralogical Record* titled *The Miguel Romero Collection of Mexican Minerals* (see figure 4.3). In the introduction to the issue, Eugene Meieran, himself one of the most prominent contemporary collectors, writes, "Since the vast majority of private collections are ultimately broken up and their cohesiveness totally lost, it is gratifying to me as a collector to see the best of the Romero collection documented in this book" (Meieran 2008:4).

The publication helps to establish the existence of the collection at one point in time, and the materiality of the book and the photographs within it act as a stand-in for the collection itself, even after the collection no longer exists. In fact, the documentation of specimens and collections in recent years, in journals such as the *Mineralogical Record* and books published by Lithographie and a few other presses, has come to play an important role in fixing the identity of collections at a given point in time and recording the possession of specimens by particular owners.

Rob Lavinsky, the dealer who handled the sale, appears to have been aware of the moral issues involved, for he worked to keep some of the pieces in public collections (the Houston Museum of Natural Science and the University of Arizona Mineral Museum) and also helped to finance the publication of *The Miguel Romero Collection of Mexican Minerals* (which was distributed free to subscribers to the *Mineralogical Record*) and provided the photographs for the book. Lavinsky took out a full-page ad at the end of the book with the heading "Finding new homes for your treasures!" printed above a photograph of the Aztec Sun and the subheading "Thank you to all my friends who have allowed me to sell their collections in recent years, including [a list of prominent collectors]." This message emphasizes the collection's continuity both through the specimens and through the relationships between collectors and dealers (as we saw in chapter 2, the line between dealer and collector is not always so clear). This allows for the persistence of ties between men (and most dealers and collectors are men)[10] and minerals accommodated by—even supported by—market exchange.

Many contemporary collectors and dealers argue that the market acts as a better steward for the future than do museums. One collector said, in a response to an appeal to donate minerals to museums, "In my opinion the market is the best caretaker of minerals, because people who will pay for them will take care of them." The visibility of minerals as they circulate from collection to collection also helps to ensure their continuity. This suggests an idea of collections as inherently more dynamic and permeable than is often assumed in the collecting literature. In the case of the Romero collection, the book served as another version of the collection that could endure even as the collection itself was dispersed.

The Romero collection's ability to assert cosmopolitan, world-class status without its Mexican-ness coming out in the wash represents a departure in the field of value-making with collections in the United States and Mexico. Unlike the Aguilar collection, it is not "put in its place" when it tries to reach across the border, and unlike the Harvard collection, it does not erase the traces of Mexican authority and agency in its constitution (though it does erase the work of miners). The collection does not choose between being Mexican and being transnational, but confidently acts as both. Its cosmopolitan, fluid character entails a fuller engagement with the market than for either the Aguilar or Harvard collections. This market-friendly stance ultimately led to its dissolution. However, the Miguel Romero collection of Mexican minerals lives on as *The Miguel Romero Collection of Mexican Minerals,* produced by the *Mineralogical Record* with Lavinsky's help.

Conclusion: Collections, Value, United States, and Mexico

Collections act as miniature worlds that hold themselves apart from but also make reference to the world outside of their boundaries (Pomian 1996). Starting from this vision of collections, this chapter has asked what kind of reference to the world outside themselves collections can make, and in what way.

Collections show especially well the two levels of value-making discussed in this book. In the negotiations through which individual pieces enter collections, value-making takes the form of "making meaningful difference." Minerals are measured according to their differences and similarities with other minerals in the collection, as a sign of their fitness for inclusion. However, these differences and similarities only make sense in the context of the second level of value-making, "making difference meaningful," which identifies and ratifies the qualities worth comparing and ranking. In the case of these collections, such qualities include things such as universality, patrimony, transnationality, and Mexican-ness, which convene and disperse in sometimes uneasy combinations. The value-making acts associated with these collections emerge out of and also help to create places—including cities such as Guanajuato and Tucson—and the connections between them.

5 MAKING PLACES IN SPACE: MINERS AND COLLECTORS IN GUANAJUATO AND TUCSON

Figure 5.1, a map featured in a report titled *Potencial Minero de Guanajuato* (Franco 1997) shows Guanajuato's centrality in Mexico, in particular the fact that over 60 percent of the country's population lives within a 350-kilometer radius (thus implying the density of infrastructure and services). The radiating circles on the map give an image of the mines and city of Guanajuato as a central origin point, with its subsoil resources expanding centrifugally to the rest of the country, and by extension, the world.

This image and the sensibility behind it contrast with many other views of the movement of mineral resources from mines to market, both for "regular people" and for social scientists. Within such views, for instance, mined ores such as silver, gold, or copper are quintessential raw materials, extracted from the "ends of the earth" and brought to the centers of global finance in New York and London. Likewise, mineral specimens are produced in geographically distant places and brought to Tucson, Munich, Denver, and other mineral marketplaces. In fact, even when minerals come from near these marketplaces, they are often treated as pristine emissaries of the margins of the cultural world. People also move from all over to a central meeting point at these mineral shows. Tucson, in particular, is called the "Mecca for mineral collectors," emphasizing its role as a pilgrimage site and meeting place for the faithful all over the planet.[1]

If we compare uses of minerals in Guanajuato with uses in Tucson, we find competing, though not totally separate, sets of value-making actions that propose alternative ways of thinking about how places are arranged and connected in space. In Guanajuato, actors use minerals to generate an expanding social world, beginning with the miner and his immediate household and spreading outward. In Tucson and other sites where min-

GUANAJUATO

POR SU UBICACION   GEOGRAFICA EL ESTADO DE GUANAJUATO PERMITE ACCESAR EN UN RADIO DE 350 km.
A MAS DEL 60% DE LA POBLACION NACIONAL.

FIGURE 5.1. "By virtue of its geographical location, the state of Guanajuato permits access to more than 60% of the national population, within a radius of 350 kilometers." Franco Ibarra, Jesús, *Potencial Minero de Guanajuato,* Dirección de Fomento Minero, Secretaria de Desarrollo Económico Sustentable del Estado de Guanajuato, 1997. Reprinted with permission.

eral collectors come together, such as the Denver-Golden-Boulder area, collectors see themselves as gathering in minerals, not only from far-flung places, but also from the margins of the social world. The multiple ways in which consumers remake minerals as *pristine* help to create an idea of central cultural space surrounded by peripheral *natural* margins and of minerals as moving from these margins inward.

A close attention to these practices suggests that the creation of value is fundamental to the creation of social and material space. Of course, these views of U.S.-Mexican space and the locations of Guanajuato and Tucson within it are not produced anew when people value minerals. Value-making occurs in a landscape stabilized through prior value-making. However, the role of value-making in making "places in space" tends to be invisible because—if value-making succeeds—the places, and the spaces in which they are arranged, seem like a priori entities. By focusing on value-making as generative, we can see more clearly the unstable, multiple, and emergent characters of places in space, as well as the presence of alternative accounts put forward by those with less access to resources or power.

The making of places in space is a kind of "project" similar to, though not quite the same as, the "scale-making projects" Anna Tsing describes in her essay "The Global Situation." For Tsing, scale-making projects are "relatively coherent bundles of ideas and practices" that engage "cultural claims about locality, regionality and globality, about stasis and circulation" (Tsing 2001:487). Tsing developed this concept in an essay calling for a critical view on the uses of *globalization* as a particular scale-making project, one that, like *modernization* itself, had become so successful that its status as one kind of project among others was no longer clearly visible. In doing so, she helped show that ideas and practices about space and scale are never unmediated, but always based on inclusions and exclusions, highlighting some things while obscuring others.

The term *project* may suggest an intentional or programmatic set of actions aimed at a consciously articulated goal. I am not using the word in this way, and in fact, I am particularly interested in *not* distinguishing between those actions that are intended to bring about a certain configuration of places in space and those that do not have this intention but do have the effect of furthering a particular configuration. I avoid the question of intention, not because I think it is unimportant, but because avoiding the distinction between intentional and unintentional action reveals the efficacy of a whole range of practices that might be otherwise hard to see.[2]

Where Tsing's interest lies with questions of scale—the local, regional, national, or global, for instance—I focus on perceived configurations of places in a larger space. How do miners in Guanajuato and collectors in Tucson imagine—and in imagining, make—their own place and the relationship of that place to larger spaces? What dynamics do they see at play between places and spaces? What vectors and links do they see as connecting places and spaces?

Fights over value—the messy process of making meaningful difference—form a key element in making and unmaking projects for producing places in space. This chapter explores the example of two rival projects put together through practices of valuing minerals in Guanajuato, Mexico, and in various places in the United States. One of these projects to produce places in space is more firmly located in Guanajuato and among miners, and the other in Tucson and among collectors. This is not a neat distinction by any means, nor are these practices and experiences unique to Guanajuato or Tucson. Neither are they stand-ins for a "view from the North" and a "view from the South." Tucson is not especially powerful, rich, or influential compared with many other places in the United States and Europe, nor is Guanajuato particularly poor or marginalized compared with other places in Mexico or in Latin America. Between Tucson and Guanajuato and between miners and collectors, however, there are significant power differentials that influence projects to make places in space. Contrasting these two projects, as related to the ways miners and collectors use minerals, helps us show the role of value-making in making places in space.

Smoothing the Way

During most of its existence, the Santa Fe Cooperative had a relatively lenient policy toward mineral specimen extraction, compared with that of other private companies within Guanajuato and elsewhere in Mexico. This changed a good deal when Great Panther arrived in 2005. In 2007, I returned to Guanajuato for a summer of fieldwork. The engineer in charge of the Guanajuato mines, Francisco Ramos, told me at that time that they were cracking down on the extraction of ore (a more serious offense) and also of mineral specimens. He said,

> In the case of those [minerals, as opposed to ore] it's not that serious, because they don't have much value, but it wastes time. A rock like this one [indicating a mineral displayed in his office, which had been confiscated from miners attempting to take it out for sale], since it's delicate, you can spend three hours taking it out. And that's half a shift.

During this time, I also spoke to former members of the cooperative. Some of those who had worked the Valenciana mine, where I had done extensive fieldwork, used the money they had gotten when they left the cooperative to rent shops in the Plaza de la Valenciana, selling trinkets and minerals to tourists and occasionally to collectors or dealers. One of these men told me,

> The company is very strict now. They don't let people take out rocks. They have guards who check, and when you come out [they say] "Let's see your backpack." But since corruption is everywhere, sometimes

there's a guard who, if you give them a little money [*una lanilla*], will give you a chance. They're always going to find some way to get them [the minerals] out. But they're scarce now.

During the same period, another person, one of the most important mineral dealers in the city, said,

Now that they've sold the mines very few [minerals] have come out. Two years ago some aguilarites were coming out that had wires, from [the mine of] Cebada. Some of them had really big wires [*alambrotas*]. They closed down the area they were coming from, but now they're going to open it up again.

In addition to being a source of income, mineral specimens in Guanajuato, and the right to extract them, also function as the raw material for social relations. The basic units of social relations constituted through minerals are ties between households. This basic model of minerals as tying together households, moreover, forms the basis for a theory of the world as centered on Guanajuato and radiating outward. In this respect, miners' uses of minerals echo Nancy Munn's description of Gawan canoes as they expand the fame of Gawa in their journeys out from the island.

In her article "Spatiotemporal Transformations of Gawan Canoes" (1977), Nancy Munn describes in detail what she called the "spatiotemporal transformations" of canoes built on the Massim island of Gawa and sent out from that island to other islands in the area as part of the Kula system of exchange (the circular movement of arm shells and necklaces most famously described by Malinowski 1984 [1922]). By spatiotemporal transformations, she means the changes, simultaneously material and laden with meaning, by which canoes are converted from an earthbound, feminine, heavy substance to fleet, oceangoing, masculine vehicles of Kula valuables and embodiments of Gawan men's greatest source of value: "fame," or "the social circulation of the self in the form of one's name" (Munn 1977:50).

Most of the transformative actions by which miners use minerals to make social worlds, unlike those performed in the production of Gawan canoes, do not materially alter the stones, because to do so would greatly diminish their price further down the line. Instead, Guanajuatense miners bring out the social character of minerals by exerting social claims over the extraction and exchange, placing them in social contexts and using them to forge and strengthen social ties.

141

Minerals as Moral Economy: The Santa Fe Cooperative

Minerals' uses within the Santa Fe Cooperative, the most important mining enterprise in Guanajuato and the holder of the city's most famous mines for much of the twentieth century, were central to their transformation into social materials. During the time of its operation, the Santa Fe Cooperative was one of the few working-class enterprises in Guanajuato that offered its workers permanent jobs as opposed to short-term contracts. This practice stemmed from its status as a producers' cooperative and its historical commitment to job preservation. Many people told me that the cooperative has a "social goal"—to operate as a source of jobs for cooperative members and their sons and grandsons (see Ferry 2005b).

This goal meant that the cooperative, among other things, maintained a much larger workforce than efficiency would demand, and that many of these workers were engaged on the surface and not underground producing ore (the only real source of cooperative income). Members often lamented to me that the proportion of surface to underground workers (*los que nos dan a comer,* "the ones that feed us") was far too high. This is also one of the reasons why cooperative wages were chronically lower than the wages of other mining companies, even when the costs of other cooperative benefits such as subsidies for medicine, schooling, and housing materials are taken into account. Although many of the older workers would not have been able to get other jobs, the cooperative still needed to provide other kinds of incentives for workers, especially younger workers, to stay. One of these was the tacit permission to extract and sell mineral specimens, something that was far more closely monitored in the private companies. Mineral selling, among other things, allowed people to continue working in the cooperative and made it one of the few viable alternatives to migration for young men in Guanajuato. It also allowed the cooperative leadership to operate on a considerably narrower margin than it otherwise could.

Moreover, the undercapitalization of the cooperative mines and the surplus of labor forced them to operate with relatively little machinery and more labor-intensive mining methods. This had the effect of causing many more minerals to be brought to light that would otherwise have been ignored or destroyed. Carl Francis, the curator of Harvard's mineralogical collection and a Guanajuato expert, said to me, "part of the appeal of Guanajuato is that you can get classic specimens at contemporary prices. Other silver localities like Kongsberg [Norway] stopped producing specimens like that centuries ago." This is due in large part to the labor-intensive and technologically simple production methods employed in the mines, especially in the cooperative.

Thus, mineral specimens were closely associated with the cooperative's social purpose—to provide employment for its members and their families, and, more generally, to keep people in the cooperative, Guanajuato, and Mexico. In the Naica mine in Chihuahua, workers sold minerals to create a fund for helping injured and ill miners and their families. Benny Fenn, a leading dealer for Northern Mexican minerals, told me how much he preferred buying minerals from Naica for this reason. In Mapimí, Durango, after the mines closed for regular production, miners formed a producers' cooperative. In 1969, the miners struck a pocket of purple adamites that was one of the biggest mineral finds of the year, with specimens that sold for prices hitherto unimaginable. This cooperative was originally formed to preserve the livelihoods of miners, who were seen as having a moral claim on the specimens.[3]

These examples highlight ways minerals can be used to construct moral economies that are wholly or partially opposed to elite or managerial notions of economic priority. This was also the case in the Santa Fe Cooperative. In the cooperative, drillers (*perforistas*) had the greatest claim over mineral specimens. This was partly because they were closest to the site of extraction and therefore did not have to leave their workplace to gather them and partly because of the greater danger and difficulty of their labor. Because drillers are closest to the blasting site and to the dust formed through drilling, they are most likely to suffer both accidents and lung ailments. The only cases in which other workers were granted greater access to minerals were when their greater needs "trumped" the default claims of drillers, such as when one worker was raising his daughter's children as his own since his son-in-law had fled the scene.

A parallel moral economy such as this is usually seen by its advocates as privileging social goals over purely economic ones (Thompson 1971). This, surely, is the case with the "traditional" rights of drillers to extract mineral specimens in the cooperative. The rights of drillers to extract stones are seen as recognition of and partial compensation for the honor and danger of work *bajo tierra* (underground). At the same time, the periodic exceptions made to this "rule" demonstrate the incipient social character of the stones. As with other moral economies described in the literature, those advocating the rights of drillers to extract stones saw their system as more responsive to the social needs of miners and more respectful of social obligations.

Practices such as the right to extract mineral specimens kept miners working for the cooperative in straitened circumstances and allowed the cooperative to continue with very low costs. In this light, the situation is more complex than a simple dichotomy between traditional cooperative organization and modern capitalist enterprise. The moral economy of the cooperative, including the rights to extract minerals, was not necessarily rendered obsolete; rather it exists in negotiation with its counterpart in "economic efficiency." The social character of mineral extraction was an integral part of these negotiations.

The extraction of mineral specimens and its associated networks of claims and obligations form the starting point of a process by which miners and other Guanajuatenses transformed the substance of the *Veta Madre*'s matrix into instruments for forging social relations among people, especially people outside one's own household. In what follows, I examine several categories of minerals' uses in Guanajuato, as commodities, as gifts, and as offerings on altars to saints and virgins. These uses exist in dialogue. Together they form a particular project that locates Guanajuato within national and global space in a particular way.

Minerals' circulation as commodities is of central concern to miners, because, as I have discussed, they can often make several times their wages selling specimens (albeit irregularly) and, in the case of the cooperative, workers sometimes remained in the mines only to have access to minerals. As with many such enterprises, the minerals market depends on the manipulation of long-term cultivation of suppliers and clients, which often overlap with other kinds of social relationships (such as biological and ritual kinship ties). Chains of supply often run among brothers, brothers-in-law, *compadres,* fathers and sons, and uncles and nephews, for instance. Two of the three *malacateros* (operators of the hoist that brings the ore to the surface) at the Valenciana had sons who operated stalls on the grounds of the mine selling minerals and trinkets. These types of exchanges speak to the importance of social ties in the commodification of mineral specimens within the city of Guanajuato and form part of those acts that transform mineral specimens into social objects. These ties, however, do not render the exchange of minerals "less economic" or "less commodified"; the circulation of minerals as commodities overlaps with and, indeed, often reinforces other kinds of social relations within

Guanajuato. In Polanyian terms, mineral commodity exchange is deeply embedded in other kinds of social patterns and practices (Polanyi 1944; see also Granovetter 1985).

Minerals as Gifts: The Fruits of Sociality

However, minerals do not only circulate in Guanajuato as commodities. Indeed, when they are sold, it is almost always to tourists, dealers, or collectors from outside. Those in Guanajuato rarely buy specimens themselves (except for resale, and then they often pay in kind rather than cash). They are much more likely to receive them as gifts, especially if they are women (Ferry 2011). For example, my friend Alicia, a cooperative member, said to me after I asked her whether she ever bought mineral specimens, "A miner might give me some stones and they have value because he took them out of the mine with his own hands." This statement strongly suggests that the value of the stone, for Alicia, comes from its ability to express the miner's labor and his journey between the mine and the surface. The mineral is nature *made social* through the miner's labor and his movement through space. In particular, the crossing between underground and surface is a journey that in itself helps to make the mineral valuable. As many anthropological studies of mines and mining have shown (Eliade 1962; Ferry 2005b; Finn 1997; Nash 1979; Taussig 1982), the boundary between surface and underground often has profound significance, marking a borderline between life and death, good and evil, human and divine, indigenous and European, and so on. In Guanajuato, in particular, as I have described elsewhere, the world of *bajo tierra* is both analogous to and also opposed to the world of *la superficie* (the surface) (Ferry 2005b:chapter 5). To bring an object from one realm to another is thus a profoundly significant act, expressed as a form of value.

From the perspective of minerals as gifts, the simile *like picking pears,* used by a miner from the mountain town of Santa Rosa de Lima about 10 kilometers out of the city, is significant. Since the nineteenth century, Santa Rosa has been populated largely by miners and by woodcutters who make firewood and charcoal.[4] Many have fruit trees in their yards or pick fruit from the nearby woods. Some poorer townspeople (usually women and children) sell the fruit at market or on the side of the road to Dolores Hidalgo that passes by Santa Rosa.

As with mineral specimen extraction, the cultivation of these fruits requires relatively little labor compared with their counterparts—agri-

FIGURE 5.2. Men dressed as *indios*—second man from left with strings of tejocotes around his neck, Santa Rosa de Lima, Guanajuato, 1997. Photo by Elizabeth Ferry.

culture, ore extraction, and processing. As with mineral specimens, the sale of fruit is often discounted in the calculation of income, though it may contribute a significant amount. Fruits from Santa Rosa, like minerals from the mines of Guanajuato, also serve as markers of place. Santa Rosa's most famous festival features strings of *tejocotes,* a local fruit, slung over the shoulder like cartridge belts (figure 5.2) and is sometimes called *La Guerra de los Tejocoteros.* Jellies and other sweets made from local quinces, peaches, and other fruits are local delicacies and are given away at parties and sold in the local restaurant. And as with minerals, hosts often give parcels of fruit to visitors. These actions transform minerals and fruit into social objects that work to bind together men and women in different households and/or families in friendly but not intimate relations.[5]

Miners and others with access to mineral specimens often give them to the secretaries, accountants, nurses, cleaning ladies, and other women with whom they come into contact. Miners also gave me specimens when I visited mines or houses; when my mother and mother-in-law came to visit Guanajuato, they gave mineral specimens to them, but not to my husband or father. One cooperative member said to me, "If I have to make a doctor's appointment or something, I bring along a stone for the secretary. It smoothes the way." This smoothness refers not only to a gift used to obtain good treatment but to making social interaction in general easier, more open and more pleasant. The image of the rough stone physically smoothing the paths between people captures the way some Guanajuatense miners see minerals as the basis for social interaction across spatial and other forms of distance.

Miners also often use minerals as offerings on altars in the mine and in their houses. These altars, dedicated to the many saints and virgins venerated in Guanajuato, thank the deities for watching over miners in their dangerous work and, in the case of domestic altars, make a connection between the house and the mine. These altars act as a nexus of social relations, bringing people together in the everyday and occasional rituals of masses, saints' days, festivals, and so on. The stones on these altars act as material markers for these social occasions and for the acts of extraction and collection that establish the mine and the miner's house as spaces dedicated to and watched over by particular divinities.

This is part of a larger Latin American Catholic pattern of placing objects on altars, which can include flowers (real or artificial), *recuerdos* (souvenirs), photos, crocheted doilies, and many other things of personal or familial significance. These form part of a general pattern of devotion that tends to emphasize the social bonds between humans and divine beings. It may seem odd to describe these offerings as specifically social acts, because they refer expressly to relations between humans and the divine. But I would argue that a feature of many strands of Catholicism (perhaps especially Mexican Catholicism) is a theory of the sacred as a world tied to the world of the profane by relations, transactions, and emotions that are akin to relations, transactions, and emotions among living humans. The food given to deceased family members in Day of the Dead celebrations in many parts of Mexico is a good example, for the ancestors are invited to the feast as if they were living relatives (Brandes 2006; León 2004; Norget 2006).

Minerals Abroad: Radiating Paths

Minerals, by their very appearance on the surface of the earth, attest to a transaction across borders (between surface and underground), and to the miner's labor in effecting this transaction. This kind of labor, as Alicia suggests, makes minerals especially valuable as social objects. Once they become objects that may cross national political borders, so some of these same associations apply. Not only are they literally chunks of Mexico that nevertheless can move long distances, but they come to light by crossing a border dense with significance and thus suggest the possibilities of other borders and other crossings. When I was about to return to the United States for a visit, my neighbor, a cooperative miner, gave me a box of minerals, labeled "Level 375/Cata Mine/Guanajuato/

Mexico" to take to my mother as a gift. This label could be seen as the verbal expression of a nested view of spatial relations, beginning with the most local (deep in the heart of the mine). The movements of the minerals from this point outward linked these spatial formations, and ultimately linked my neighbor in the mountain town of Santa Rosa de Lima with my mother in Cambridge, Massachusetts (Ferry 2005b:187). This is the kind of practice fueled by the generative labor of miners that smoothes the way between places and sends the mineral substances from Guanajuato to the rest of the world. It also implicitly assumes Guanajuato itself as the source and center of these global radiations.

Not all paths are equally easy to make smooth. Just before he gave me the box, my neighbor stopped and asked, "But are you going to have trouble getting them across the border? Will you have to pay customs?" This is a perfectly reasonable question, especially given that minerals are subsoil resources (though rarely subjected to the same regulation as ore). At the same time, of the many things people gave me to take home, it was the only time this question arose (except when people satirically suggested climbing into my suitcases themselves, which often happened). It seems possible to imagine that minerals, like people, are potential border-crossers and thus might potentially be prevented from crossing. Minerals from Guanajuatense mines, the productive practices that bring them to the surface, and the local webs of meaning-making in which they are enmeshed thus make them apt materials for the construction of a social world, paving stones for the smooth paths that link people, households, and places together.

In most cases, these forms of exchange (as gifts among people and between people and deities) are conceived as *local,* that is, between households of Guanajuatenses or between Guanajuatenses and their saints, who are viewed as intimate friends. But they can also serve as a model for more long-distance forms of sociality. Minerals, like silver and other products of the mines, are rooted in Guanajuato and yet have the capacity to travel the world over. The transformational acts by which "naturally" rough and motionless objects begin to move and to make other things smooth are themselves a form of space-making, rooted in Guanajuato and radiating out to the rest of the world.

This notion occurs in other contexts as well. Silver ore is also often seen as a substance that leaves Guanajuato but returns (or should return) as other kinds of things, such as children, houses, and public and religious buildings (Ferry 2005b:chapter 6). The image from the report that I placed at the beginning of this chapter (figure 5.1) gives a visual representation of this idea. Although mineral specimens are not considered patrimony in the same way that silver is, the notion of Guanajuato as a place from which substances travel along value-making paths is similar. This view of Guanajuato as a central place from which marvelous mined substances radiate and return runs alongside more powerful views of Guanajuato and places like it as "marginalized" producers of primary commodities for metropolitan centers.

The notion of Guanajuato as the center of the country is buttressed by local historiography, which emphasizes that the geographic center is in nearby La Luz, marked by the statue of *Cristo Rey* (Christ the King). Furthermore, Guanajuato lies within the Bajío region, which plays a significant ideological role in the formation of the Mexican nation. It is the region that was most economically developed in the colonial period, with the engine of the economy in the eighteenth century coming from the mines of Guanajuato. Indeed, it was the vitality and diversification of the economy surrounding Guanajuato that made Ángel Palerm describe the Bajío as "the first world economic system" (1980; see also Tutino 2011). The movement for independence from Spain also began in the state of Guanajuato, in Dolores Hidalgo, and the first major battle against the Spanish took place in Guanajuato cities. Indeed, the hero of the battle was a miner named *El Pípila;* his statue teeters on the Panoramic Highway above the city. These events earned Guanajuato the title "Cradle of Independence," a moniker that emphasizes the city and region as generative source.

It is extremely common for cities and regions to emphasize their central role in events of national importance such as the War of Independence, especially in state-sponsored forms of history such as textbooks and civic monikers. And, arguably, many places in the world regard themselves as "central" in some ways and at some moments. The particular character of Guanajuato's centrality, as some see it, lies in the role that mined substances such as silver and minerals play in locating Guanajuato and binding it to the rest of the world. The outward radiation of minerals and silver in the world and the subsequent enrichment and civilizing of Guanajuato make the mines the primary engines for Guanajuato's centrality and constitute a dynamic vision of the expansion of the "fame of Guanajuato," to modify the phrase Nancy Munn applies to the movements of Gawan canoes from Gawa to other islands (Ferry 2005b:178; Munn 1986).

In sum, Guanajuatenses in many contexts certainly do ascribe to the view of their city as bound to the rest of the world through political and economic relationships that keep it in a relatively subaltern position within global capitalism. Indeed, it is for this reason that Guanajuatense miners played a significant role in the drafting of the new national constitution after the Mexican Revolution. The Santa Fe Cooperative was founded as an attempt to Mexicanize the mines, to keep value in Guanajuato, and to fight against the marginality of Mexican mining in the world economy. At the same time, a minority report within Guanajuato does not emphasize the model of Guanajuato and its mines as marginal or peripheral, but poses Guanajuato itself as the center of the world, or at least the world forged through the movement of its silver, gold, and minerals, which comes to nearly the same thing.

Knowing what we already know about mineral sciences and mineral collecting in Mexico, it is clear that not everyone in Guanajuato uses minerals in quite this way. For one thing, the differences between the ways people in Guanajuato and in the United States use minerals are full of slippages and qualifications. For example, some miners' homes have mineral assemblages surrounding the image of a saint or virgin, and I am calling these *altars*. Others have more secular groupings, arranged either as decoration, memorabilia, or more scientific collections. Many miners and miners' wives reported that they simply liked to have minerals in the home because they were pretty and unusual. One former miner had a collection laid out much like a collector; rather than arranging the stones in a picturesque configuration, he had them in rows and columns as they might be in a museum. When I asked him about this, he said, "There are guys in Las Torres and Peregrina who make a business out of them. I don't know anyone else who collects them like I do." Class position plays a role in this. Mining engineers in Guanajuato also have assemblages of minerals in their offices and homes, but these were much more likely to make reference to scientific, aesthetic, and occupational associations than to the social labor and social relations represented by minerals. And, as we saw in chapter 4, collectors like Ponciano Aguilar and Miguel Romero use minerals in very different ways than do many miners.

In the midst of this variety of uses, however, we can discern a pattern of use that expresses a particular spatial arrangement that depicts Guanajuato as the core, not the periphery or semi-periphery. Minerals, as nature made social through human labor, can help to "smooth the way" in these global radiations. Once these minerals cross national and class borders, however, other forms of labor make them into other kinds of valuable objects and contribute to other projects to make places in space.

Cultivating the Pristine

At any of the major gem and mineral shows in Europe and the United States, you will find the "booths" of the company Collector's Edge Minerals. I use quotation marks here because Collector's Edge pays a lot of attention to the design of the area they rent on a show floor, evoking a museum or high-end jewelry store (e.g., Tiffany's) rather than a commercial venture usually evoked by the word *booth*. The space is carpeted in a discreet color, with comfortable modern design chairs and tables arranged in casual groups. Glass cases lined in black velvet contain artfully arranged and lighted displays of exquisite minerals.[6] The pieces are mounted on custom-made clear acrylic mounts with the species name and locality etched onto the front.[7]

To use Stephen Greenblatt's description, the minerals displayed in this way are intended to evoke "*wonder* . . . the power to stop the viewer in his or her tracks, to convey an arresting sense of uniqueness" rather than "*resonance* . . . the power to evoke in the viewer the complex, dynamic cultural forces from which it has emerged and for which it may be taken to stand" (Greenblatt 1991:42). Put slightly differently, their power is meant to be iconic, in that they are to be taken as qualisigns of beauty, luster, and *pristineness* rather than indexical, as pointers to and examples of particular mineral species or a particular locality.

The Collector's Edge showrooms are generally full of customers and window shoppers, as well as collectors and other members of the mineral world catching up on news, passing the items, and swapping information. Collector's Edge is a leader among high-end dealers and a highlight of the shows. In September 2005, I interviewed Steve Smale, one of the world's most famous mineral collectors (and also a renowned mathematician), in the Collector's Edge showroom. We sat on a sofa in the middle of the room, while collectors and shoppers milled about us, casting curious glances and chipping in comments from time to time. Steve has no formal connection to Collector's Edge, but he is a supremely valued client. Moreover, because he is known as having impeccable taste and judgment in minerals, the company is pleased to have him be seen occupying their rooms.

We discussed the different criteria that might make a mineral beautiful (or *aesthetic,* the word used by Smale and in general by such collectors), and he emphasized the subjectivity of such criteria (though as with all judgments that appear entirely subjective, a collective set of norms and forms has emerged). He emphasized such dimensions as the *integrity* of the piece, or its overall composition, including the relationship between the main mineral and its matrix, and its *horizon,* the relation between the edge and surrounding space as seen from eye level.

As we saw in chapter 2, valuing minerals as aesthetic objects has greatly increased over the last thirty years, particularly at the high end of the market. However, the minerals are valued as not just any sort of art, but as *natural art.* They are not made by any human hand, but by the interaction of the forces of nature: chemical bonds, conditions of temperature and humidity, and so on. In fact, this is much of their appeal, according to many. Dealers and collectors sometimes contrast their liking for minerals with that of lapidarists, who purchase or collect "rough" stone to be cut and polished into gemstones and jewelry. In an online discussion between lapidarists and mineral collectors, mineral collectors made the following remarks:

> There is nothing that can rival a beautiful natural mineral. That is not to put down those who cut, polish and mount. It is an art and it [is] beautiful. Just that some things are "beautifuller."
>
> I am a chemical engineer. When I admire a specimen, I have some idea of the many forces and laws of physics and chemistry that have interacted to produce that one sample. But, I am awed by the fact that they combined to produce such beauty. Further, such beauty of structure and color and combination is far superior to anything I could invent. (Rockhounds Forum 2002)

Similarly, the mineral dealers Stuart and Donna Wilensky proclaim on their website:

> Every mineral specimen you see on our web site is completely natural and as found deep in our Earth. The only intervention by human hands is the removal from the mine, cleaning and sometimes trimming away parts which are not desirable. (Wilensky 2009)

These excerpts capture some of the important features of naturalness as a marker of value for mineral collectors. These include *pristineness,* or the idea that the mineral is untouched or virtually so, and the notion of the mineral as the product of numerous unseen forces and laws combined in unique and serendipitous fashion.

The dealers and collectors who inhabit and create this world are not *producing* minerals in the sense of extracting them from the mine, but they engage in various kinds of labor that make minerals into new kinds of things. In general, such productive labor is aimed at remaking minerals into natural and pristine objects. Through a variety of words and deeds, dealers and collectors work to render the naturalness of mineral specimens visible and effective. Thus, in their circulation and use in the United States, the value of mineral specimens depends on the erasure of the very kind of social transactions and associations that were highlighted and celebrated among miners in Guanajuato. We can see this especially clearly if we contrast the Wilenskys' description of their minerals with Alicia's statement. Alicia values those minerals that have been taken from the mine with the miner's "own hands," whereas for the Wilenskys, the less intervention from human hands, the better.

Forging the pristine in this way takes a lot of work, and a number of specialized techniques have been developed to carry out that work. These include preparation, photography, and narrative description.

Preparation: "Collectors Don't Like Smooth Edges"

Preparation refers to the process by which the dealer or collector cleans a specimen and removes "excess" material that detracts from its beauty. It creates a specimen that appears naturally striking and proportional, without leaving any indication of the human labor that went into its creation as such. Many dealers and collectors do this work themselves, but for the most expensive minerals, there are mineral preparation laboratories. The first of these was the laboratory operated by Collector's Edge at their main offices in Golden, Colorado.

In September 2005, I visited this laboratory and was given a tour. The lab itself looked like a high-tech lab, with large tables gleaming with black, white, and silver instruments (many of which, I was told, were adapted from dentistry), sinks, expansive shelving, fluorescent lighting, and a hushed, serious atmosphere. A few large, bright minerals were placed on tables, seemingly in the midst of preparation. My guide described the process, which begins with removing any sign of the pick, drill, or other implement by which the mineral was removed from the mine. Then the area where this work has been done is roughened, because "collectors don't like smooth edges." The later stages of the work involve much more fine-grained work and careful decision-making. They may also involve careful repair of specimens. Such repair must be disclosed to the buyer, and often brings down the price of a specimen. However, the existence of new techniques to repair specimens, and a new culture of acceptance of certain kinds of repair, has definitely raised the ceiling on the size and quality of minerals on the market. I asked my guide how long it typically takes to prepare a mineral in the lab. He responded, "It can take a couple of hours or it can take a year. And the end result is what you see in the showroom—a really *pristine* piece."

This use of the word "pristine" to describe the end result of a process rather than its starting point is not only an almost absurdly good example of Marx's fetishism of commodities. It also expresses how U.S. mineral collectors construct the places that minerals come from as virgin territory, legible mostly as sites of potential natural resources and natural art. The valuing of minerals as pristine pieces of the earth presumes a system of global relationships in which value is extracted from all over, but only recognized—and thus realized—by those in the United States and Europe.

Preparation can be controversial, as it includes such things as the repairing of broken specimens, which was considered unethical until the late 1960s. According to Dave Wilber, the renowned dealer and collector linked to the aesthetic minerals movement (chapter 2), the ice was broken on allowing repairs around 1969 when an advanced collector displayed a phosphophyllite (a rare zinc phosphate mineral of a delicate blue-green color) and then said, "what would you say if I told you it was repaired?" Now repairs are considered ethically neutral as long as the repair is reported (and a repaired specimen will be priced below an equivalent nonrepaired one).

Controversy over what constitutes legitimate preparation and what does not reveals the importance of the mineral as unmediated nature: repairing is acceptable because it is restores the mineral to its natural state, what it was "supposed to" look like. Illegitimate restoration—fakery—involves enhancing a mineral beyond what it might have looked like at any point, for instance by heating it to intensify its color or changing the crystal form through microabrasion.[8]

Mineral Photography: The Floating Icon

Mineral photography also works to transform mineral specimens into pristine natural objects that display aesthetic properties without reference

FIGURE 5.3 Pyromorphite 7.2 cm, Roughton Gill mine, Cumbria, England, Ex–Charles Jones collection, bequeathed to Heidelberg College, Tiffin, Ohio, in 1919. Collection of Karl Warning, Dallas, Texas, U.S. © Jeff Scovil. Reprinted with permission.

to human labor. Techniques of placement, lighting, color, and exposure are intended to reveal the geological and aesthetic properties of the specimen in isolation, enhancing the stone's iconic function as representing qualities of color, luster, and form that are the product of nature rather than human labor (though, as should be clear by now, quite a lot of labor has gone into producing this effect).

Jeffrey Scovil, based in Phoenix, Arizona, is probably the world's most famous and successful mineral photographer. In an interview in August

2006 at the East Coast Gem and Mineral Show, he described to me his entry into the field of mineral photography. As we talked, he photographed specimens for collectors. A young man, the assistant to one of the dealers, came and asked to watch his technique. Jeff set up the minerals on a glass, secured them with bits of putty, and placed colored placards behind and below them to bring out the colors of various pieces. The final effect of these photographs is the absence of shadow and the general sense that the minerals are floating in space (figure 5.3).

He said:

> There didn't used to be mineral photography as such. There's a reference in a book by Gratacap in 1915, and there's Fred Pough's work in the 1950s with the American Museum of Natural History, but it wasn't much. They used to take the photographs on wide wale corduroy, that was when curators used to stick the catalogue number right on the front of the mineral, instead of somewhere you couldn't see it. The real beginning was with Lee Boltin in Paul Desautels's *The Mineral Kingdom.* I got that book for my high school graduation—and I couldn't stop looking at it.

This statement shows the way in which mineral photography came to detach itself from the general field of scientific or natural history illustration and to become allied with (and to promote) the development of aesthetic appreciation of minerals. These days, as Jeff described it to me, every major collector has his collection photographed; as he said: "If you get a photo in *Min Record*—or the cover—it's an immediate increase in the value of the specimen."

By "Min Record," Scovil meant the magazine *Mineralogical Record,* the magazine for mineral connoisseurs, which is published quarterly in Tucson, Arizona. The *Mineralogical Record* has the capacity to ratify both the value of a specimen and its links to a particular dealer or collector. One collector and "mineral broker," who works helping to arrange and represent exchanges between the most elite collectors (I will call him George Guss), brought me to the *Mineralogical Record*'s booth at the Colorado Gem and Mineral Show and showed me the photograph of a mineral captioned "ex-George Guss collection." "Now I can pass it on," he said, "since my name is attached to it."[9] Sometimes mineral photographs can document a whole lineage of owners, as in the case of a poster on display at a leading dealer's showroom at the 2005 Denver Show. It showed a magnificent blue topaz and was signed by the previous four owners and the photographer (figure 5.4).

156

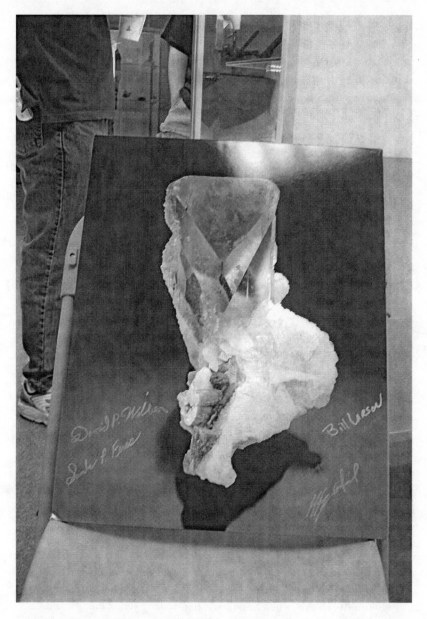

FIGURE 5.4. Poster of blue topaz, Denver Gem and Mineral Show.
Photo by Elizabeth Ferry.

Mineral photography's lingering gaze on the mineral as a sumptuous object that appears isolated from all contact (even with the surface on which it rests), and the presence of venues for such photography play a major role in promoting minerals as icons of beauty and pristineness.

Narrative Description: Nature's Design

Narrative descriptions in magazines, on websites, and other places also enhance minerals' pristineness. Stuart Wilensky writes on his website, for instance:

> I (Stuart) grew up in a very artistic family. My parents were art and antique dealers and I traveled the world with them visiting museums and studying every period of art from Egyptian to 20th century. In college I earned my masters degree in art history. Little did I know that every artist, every creative work I had ever seen would pale in comparison to the work of Mother Nature. My lifetime of aesthetic learning drew me rapidly towards the natural beauty of minerals. It was as if a lifetime studying art had brought me to this point where I now understood that all art, all aesthetics, come from nature. The appreciation of human art is but a reflection of nature itself. (Wilensky 2009)

Here the privileging of natural over human art is based on an almost theological view of the world, in which human design tries to reproduce the perfect world of nature. This sense is often present in descriptions of the discovery of famous minerals that draw on a sense of predestination. Such a sense allows minerals to be removed from the earth without losing their pristineness.

For instance, in October 1981, a pocket of adamite crystals with rare purple terminations was found in the San Judas vein, level 6. Mike New, in the Mapimí issue of the *Mineralogical Record,* described the moment of seeing the rocks for the first time,

> On October 24, 1981, my 36th birthday, we started working a very hard bit of *fierro* [iron ore]. Pedro was hand-drilling, and suddenly his chisel sank in full length; when he pulled it out, water gurgled out of the hole. . . . When we pulled a large piece of fierro away from the working face, we found ourselves staring into the prettiest pocket I had ever seen. There were about 10 specimens still attached to the walls, with adamite crystals up to 6.25 cm. The small toenail-sized specimen pictured on the cover of the *Mineralogical Record* . . . was lying on the floor of the pocket, just below its former point of attachment. The piece that was

later sold to the Sorbonne was in the back of the pocket, and the best piece, though smaller, was lying loose on the floor. (New 2003:50)

New's description of the movement of stones from fortuitous discovery to final enshrinement in famous magazine covers and collections is characteristic of specimen biographies as told by collectors and dealers. The passage addresses a like-minded audience of mineral lovers (those who know the specimens in their later lives), asking them to imagine the moment when these now famous specimens were first seen by human eyes. It thus creates a sense of predestination or foregone conclusion, what we might call *nature's design*. Nature's design combines the orderliness of aesthetics (balance, purity of color and form, and so on) with the precarious untouched quality that has become so highly prized (see Walsh 2010 for a similar process in the marketing of natural sapphires).

This narrative seeks to represent the value of the purple adamite specimens from the San Judas pocket as already present when the minerals are still in the mine, waiting to be discovered and destined for certain glorious ends (the cover of the *Mineralogical Record,* the Sorbonne). However, the value must be recognized by the right person, in this case Mike New, a prominent specimen miner and dealer, and holder of the specimen rights to the Ojuela for much of the last few decades. The fact that New discovered the pocket on his birthday makes manifest his status as chosen discoverer.

Oddly, the forging of pristineness gives a certain sameness to all minerals, regardless of where they come from. Certainly their particularities matter, and their locality and provenance can affect their price. But as icons (signs that represent a particular quality) rather than indices (pointers to a particular species or locality), they signify an undifferentiated realm of "the pristine," strongly influenced by Romantic poetry and painting (Steve Smale's interest in the horizon of a piece, for instance, evokes the English Romantic or Hudson River School painters).

In the previous section of this chapter, I argued that miners launch a project to produce their place in space that is similar to the Gawans described by Nancy Munn, with the movement of minerals radiating value outward from Guanajuato. In contrast, the high-end dealers and collectors draw on an iteration of space like that discussed by Mary Helms in her book *Ulysses' Sail,* an extended meditation on the symbolic power of geographical distance across a range of cultural contexts (1998).

This geographical distance need not be measured in kilometers, and for that matter, the category of the pristine created by mineral dealers and collectors makes all minerals equally distant from the center of human, social life.

Helms writes,

> In traditional cosmologies [like that of early twenty-first century, middle-class, Anglo, male mineral dealers and collectors] places and peoples located at a distance from a central heartland have been conceptualized as being increasingly different from or contrastive with the central axis mundi. (1998:262)

In the case of minerals, the contrast is between the fully human, social world of the United States and the nonhuman (geological) world of the pristine sources of minerals, wherever those might actually be located. Helms further remarks that one important category of symbolically powerful objects from afar is "valuable raw materials . . . derived unformed and unshaped (uncrafted) from the earth, air or sea and which are therefore redolent with the power of the universe" (1998:115). Minerals would fall squarely into this category. The dealers who bring these objects in from afar become themselves to be seen as imbued with power, as do the collectors who show their ability to possess objects coming from the pristine realm of faraway nature.

Conclusion

In the accounts I have laid out here, we see two different visions of spatial relationships, traversed and bound together by minerals. One, which tends to be articulated by miners in Guanajuato and their families and friends, produces value by bringing minerals across the boundary between *bajo tierra* and *superficie,* giving them as gifts, circulating them in systems of moral economy, communicating with the divine, and so on. These value-making acts underscore a view of Guanajuato as a central place from which value emanates, creating a social world linked by "smooth" paths.

Another view, one that tends to be linked to elite dealers and collectors in the United States, remakes minerals into pristine objects of aesthetic appreciation. Here minerals are no less social in their function than in Guanajuato, but the work of making them valuable focuses on their perceived naturalness. Here the center of the world appears to be the United States (or Europe), where connoisseurs recognize and appropriate the value that comes from outside. Mexico appears as a "rough," pristine land, waiting for proper forms of exploitation and evaluation.

At the same time, it is the social networks of extraction and circulation originating in Guanajuato and other mineral localities that allow dealers to get their hands on the stones. For this reason, even as they erase these connections and the human actions that brought the stones from the mines, mineral dealers and collectors depend on them. Moreover, mineral dealers and collectors also use rocks to expand their own social connections. Think of the blue topaz poster (figure 5.4) signed by its four most recent owners and by the photographer.[10] It is not that minerals are no longer social objects once they cross the border. But their actual social power derives from their ability to seem like icons of pristine nature.[11]

In his article "Cosmologies of Capitalism: The Trans-Pacific Sector of the World System," Marshall Sahlins documents several different cultural logics that operated in the early decades of European contact and trade in the Pacific. He does so to counter what he saw as a tendency in anthropological accounts of colonialism and capitalism to oversimplify the process by which these forces integrated different places into a world system. He writes, "the World System is not a physics of proportionate relationships between economic 'impacts' and cultural 'reactions'. The specific effects of the global-material forces depend on the various ways they are mediated in local cultural schemes" (1988:5). In a similar vein, Nicholas Thomas (1991) juxtaposes the "Indigenous Appropriation of European Things" and the "European Appropriation of Indigenous Things" in his treatment of exchange and material culture in the Pacific Islands, as a way of emphasizing their analogous character.

The case under discussion is a little different, because Guanajuato, Tucson, and Denver are equally the products of European capitalism. Guanajuato, after all, was founded as a mining camp in the sixteenth century in an area previously within the territory of nomadic groups, very few of whom settled in the area. The mines attracted Europeans, indigenous people from farther south, and (a few) Afro-Mexicans, who combined to form, in the words of one historian, "a society that constituted itself as something new" (Guevara Sangínes 2001:170). Although there may be some precapitalist cultural influences at play in Guanajuato, they are no more present (in fact, perhaps somewhat less so) than they would be in Tucson, also a new society formed of the interaction of Europeans and indigenous people brought together in capitalist relations.

However, miners in Guanajuato and collectors in Tucson and Denver are very differently positioned within networks of commodity extraction, and therefore they have perspectives that cannot be reduced to one another. In this sense, Sahlins's and Thomas's arguments help to show us how different places confer meaning onto the transactions of valuable goods. Looking at Guanajuato miners' and Tucson collectors' perspectives side by side helps us to realize that the view of value flowing inexorably from South to North is not the only possible view nor the only possible arrangement. These different projects do not operate on an equal playing field. Like "globalization," the view of the United States as a center that brings value toward it from other places "draws us inside its rhetoric until we take its claims for true descriptions" (Tsing 2001:330), though less so than even a few years ago. In the terms I used in the introduction to this book, this view has more success in establishing itself as a black box that can then be used without looking inside—almost like a $5 bill.

In Mexico, as in many other places in the formerly colonized world, notions of the place as a source of natural resources and a home of natural persons (*naturales* was the colonial Spanish term for indigenous people) have played an important role in naturalizing relations of power and inequality between colonizers and the colonized. This process is well established and thoroughly (not to say exhaustively) discussed in the anthropological literature (for two excellent examples, see Lutz and Collins 1993 and Price 2007). However, if Mexican minerals spring parthenogenetically from the earth and are not supposed to be changed (as the view promoted by some U.S. collectors suggests), we are left with other questions: Why are minerals so much more expensive once they reach the high-end dealers and mineral shows than they were in Mexico (more than transportation and other costs could account for)? What can dealers and collectors claim as the value added if they are also committed to the minerals' pristineness?

MINERAL MARKETPLACES, ARBITRAGE, AND THE PRODUCTION OF DIFFERENCE

In November 1977, a prospector named Felix Esquivel entered the Ojuela mine in Mapimí, Durango, Mexico, and discovered a pocket of over twenty specimens of legrandite (a zinc arsenate). He and his brother sold them to a dealer, Jack Amsbury, via his Mexican agent Shorty Bonilla, for 48,000 old pesos, or approximately $4,000. Amsbury, with the help of Gene Schlepp, a Tucson dealer, resold the best piece (later christened the Aztec Sun) from the pocket (figure 6.1) to Miguel Romero for $30,000. Felix Esquivel reported to us that the buyer (he did not remember who) did periodically pay him 1,000 new pesos—between $60 and $100, depending on the year—as further remuneration. Romero held onto it for the rest of his life, planning to donate it with the rest of his collection to the University of Arizona Mineral Museum, where the best of his collection was on loan for many years (see chapter 4). In 2008, his heirs sold it to a Lebanese collector for something in the neighborhood of $1.7 million.

The vast sums made on this one piece emerge through a process of *arbitrage:* that is, profit made by taking advantage of price differences for the same commodity in different markets. The opportunity for arbitrage creates markets for mineral specimens—if the chance to make a fantastic profit by buying a mineral at its source and reselling it in Tucson or another marketplace did not exist, the mineral economy would be a fraction of what it is.

In the financial world, arbitrage often refers to the nearly simultaneous purchase and sale of financial commodities (securities, foreign exchange) to take advantage of price discrepancies. However, the term has been used more broadly to refer to things like maneuvering between different regulatory contexts in banking and elsewhere (regulatory arbitrage), offshore production to take advantage of a cheaper labor market

FIGURE 6.1. The Aztec Sun legrandite specimen, Ojuela mine, Mapimí, Durango, Mexico. © Jeff Scovil. Reprinted with permission.

(geographic arbitrage), or the appellation d'origine contrôlée system (cultural arbitrage) (Ghemawat 2003).

Much of this literature assumes difference solely as a precondition rather than an effect of arbitrage.[1] And in the cases of labor and capital costs, it also assumes that this difference is likely to disappear. Indeed, economic theory argues that arbitrage undercuts the conditions of its own existence (for a discussion of the temporal implications of this, see Miyazaki 2003). The more people practice arbitrage, the smaller the gap between markets that makes arbitrage possible is supposed to become. The formation of mineral markets in Tucson and Mapimí not only depends on the ongoing production of difference (among places, people, and objects), but also that the very practices of arbitrage between these markets help to create difference. At least sometimes, difference can be an effect of arbitrage as well as a cause.

The Mapimí market has been around for over fifty years, and opportunities for arbitrage are not going away. This is evident to anyone who goes to the two places, but actually quite hard to measure, for two reasons. First, some dealers enter into long-term financial relationships with suppliers so that it is difficult to bound and compare particular transactions. Second, dealers typically will not tell people how much they spent for a particular piece and to ask would court distrust (this taboo against asking is a strategy that helps to keep the price gap open).[2] However, the profit realized on the above-mentioned legrandites is high but not preposterous, particularly for a new and unexpected find. Moreover, though prices in Mapimí are going up, they are not rising so much as to force the arbitrage gap to close.

Because Tucson is only a one or two days' drive from Mapimí and many Mexican miners and local mineral dealers have crossed the border multiple times (for instance, to work as carpet-layers in Denver and elsewhere), transportation costs and political boundaries cannot fully explain this situation. These markets' heavy dependence on arbitrage would suggest that the gap in prices would close as economic equilibrium is reached. Why, then, does the gap persist over time? One answer lies in the uneven access to social and cultural capital of miners and dealers in Mapimí and dealers and collectors in Tucson.

In his 1986 essay "The Forms of Capital," Pierre Bourdieu describes *capital* as:

> what makes the games of society—not least, the economic game—
> something other than simple games of chance offering at every moment
> the possibility of a miracle. . . . [Capital] is a force inscribed in the
> objectivity of things so that everything is not equally possible or impos-
> sible. And the structure of the distribution of the different types and
> subtypes of capital at a given moment in time represents the immanent
> structure of the social world, i.e., the set of constraints, inscribed in the
> very reality of that world, which govern its functioning in a durable
> way, determining the chances of success for practices. (p. 241)

Bourdieu goes on to identify "three fundamental guises" in which capital can present itself: economic, cultural, and social. Cultural capital can appear as *embodied* (language, bodily disposition, accent, taste, etc.), *objectified* (culturally prestigious objects, such as a mineral collection), or *institutionalized* (academic degrees, membership in prestigious societies, union membership, etc.). Social capital refers to the "the aggregate of the actual or potential resources which are linked to possession of a durable network of more or less institutionalized relationships of mutual acquaintance and recognition" (p. 244) as expressed in the adage "It's not what you know, it's who you know." Even though this typology has been critiqued for its "Borg-like conception of capital" that ignores contingency, agency, and the dialectical relationship between capital and labor (Rikowski 2008), it does help to explain why some people in the mineral economy can buy cheaply and sell dearly while others cannot. But this answer immediately yields to a further question: How is the social and cultural capital needed to participate in the higher-priced Tucson market distributed to some and not to others?

The persistent gap in prices in these two marketplaces is especially perplexing because of the specific ways in which high-end collectors value minerals. As we saw in chapter 5, dealers and collectors value minerals in large part for their apparent pristineness; minerals that bear visible traces of extraction are far less desirable and any repair or restoration must be disclosed. And yet this pristineness is highly cultivated. Careful preparation of minerals seeks to erase all traces of mining, and dealers describe minerals as nearly "untouched by human hands."

If extraction and subsequent distribution are not supposed to alter mineral specimens (regardless of whether or not they are in fact altered), what justifies these huge price differences? Where diamonds and other gems claim to have at least some value added from gem cutters in Antwerp and their hundreds of years of craftsmanship, no such explanation will work for minerals. This tension between the need for pristineness and the steeply climbing price structure means that the kinds of labor that make value in ways that do not alter the mineral (or at least do not appear to do so) are even more important. Much of this labor, I hope to show, focuses on the distribution of social and cultural capital that gives different actors in this process access to minerals in different marketplaces.

Labor to distribute social and cultural capital clusters differently in different markets. In Tucson, labor clusters around the making of the connoisseur and his ability to recognize the stone's value. In Mapimí, it clusters around the making of the miner with rights to the mineral and networks through which to distribute it. The first operates within an ideology of connoisseurship; the second within one of rights to access. Both use complex strategies to regulate membership in a class of persons able to buy and sell in each marketplace. Both aim to define the kind of person who can lay claim to the mineral at different points in its journey, in ways that are strongly inflected by class, race, nation, and gender. These attempts are never completely successful, but they are successful enough to keep open the window for arbitrage, upon which these markets depend.

This kind of practice is by no means unique to Tucson and Mapimí or to minerals. Indeed, the analysis I present here can help to explain many situations in which markets remain distinct in spite of (indeed, in some ways because of) frequent transactions between them.

Tucson

Strategies to regulate access to minerals in Tucson (and specifically access to those collectors willing to pay the highest prices) tend to focus on the cultivation of knowledge and connoisseurship, insertion into appropriate social networks, and the ability to speak English and to manifest a middle-class habitus.

Making Connoisseurs: Mineralogical Record

One major site for these practices is the *Mineralogical Record,* the premier U.S. magazine for mineral collectors, published in Tucson, with a circulation of approximately 5,500 subscribers. The *Mineralogical Record* helps to cultivate the cultural and social capital needed to be a mineral collector (especially a high-end collector) or to be able to sell minerals to these people. It does so in large part by addressing its readers as mineral connoisseurs, while giving them the knowledge that is seen as constituting connoisseurship.

The magazine provides material necessary to develop these qualities, including articles on mineral localities, famous collectors and collections, and mineral species. Often the issue will be devoted to a particular mineral locality (e.g., Erongo [Namibia], Llallagua [Bolivia]) or to a country rich in localities (e.g., Madagascar, Mexico). The March/April 2010 issue includes the following articles: two articles on sperrylite (a platinum arsenide) from two different localities (Talnakh orefield, Siberia; Tweefontein Farm, South Africa); an article on the Fanwood quarry, New Jersey; an article on the Brown Monster and Reward mines, Inyo County, California; several obituaries of people in the mineral collecting world; a note on "the neurology of collecting"; and a note on the Victorian four-sided mineral cabinet. This range of topics is fairly typical, though historical articles, show reports, and other features also appear regularly.

The articles are illustrated with stunning photographs of mineral specimens as illustrations to articles and show reports and in advertisements. The magazine itself is beautifully produced and has won several Gold Ink Awards, a prestigious annual prize cosponsored by printing and graphic arts industry magazines such as *Printing Impressions* and *Publishing Executive*.[3] Since 2005, the cover photo for every issue has appeared against a white background, with the magazine's title in a copperplate-like font. This design gives a classic, slightly scholarly but still luxurious look to the publication.

The design choices help to establish a tone for the magazine and the qualities of connoisseurship it exemplifies, and these are repeated in many of the articles. The locality articles, which are the most common, include historical information (complete with period photographs or drawings), mineralogical and crystallographic data, and spectacular photos of specimens from that locality (with accompanying provenance). In a guide to writing locality articles posted on the website, the editor-in-chief writes, in reference to specimen photographs, "Like Oscar Wilde, we at the *Mineralogical Record* and our faithful readers have the simplest of tastes: we are easily satisfied with the best" (Wilson n.d.).

This statement whimsically but aptly captures the magazine's sense of quality, what goes into it, and what role the magazine plays in judging it.

The *Mineralogical Record*'s design and content seek to create a mineral consumer with the appropriate knowledge and taste for the Tucson market (and other collectors' markets around the world). This is clearly demonstrated in the prefaces to a series of special issues on Mexico, each of which ends with a (tongue-in-cheek) salutation to the reader, such as the following:

> So sit back now in that comfortable leather chair, prop your silver-studded boots up on the mesquite coffee table . . . and enjoy issue #1, on the minerals of Boléo [Mexico]. . . . *Adelante!* (*Mineralogical Record* 1998:3)

Have your significant other pour you a small glass of well-chilled, high-quality sipping tequila (one could do worse than Corazón Añejo from the Arandas highlands of Jalisco: 100% blue agave, aged for two years in charred oak casks), just to get in the mood. Soon you will be hearing the distant howl of the Mexican wolf, the muffled boom of carefully placed explosive charges, the excited cries of the miners as they break into yet another fabulous pocket, and the rhythmic rumbles of the old hoist engines as they bring more treasures to the surface. May that special specimen you have been looking for be among them. (*Mineralogical Record* 2003b:3)

Make your way back to that leather easy chair (the one with the steer horns on the back). Stoke the stone fireplace with a few more mesquite logs to keep it burning brightly. Pour yourself a small crystal-glass of sipping tequila. Then warn the other *mineros* and *barreteros* (miners and drillers) not to bother you for a while, as you settle back to enjoy another vicarious journey to some of Mexico's most famous and most productive mineral localities. (*Mineralogical Record* 2004:3)

What kind of reader is called into being through these passages? He is, I would say, presumed to be male and from the United States, surrounded by objects evocative either of the western United States or of northern Mexico. His knowledge and love of minerals go hand in hand with his taste for other forms of what Bourdieu would call "objectified cultural capital" (1986:47). The narrated experience of reading draws on cinematic evocations of Old Mexico, finally arriving, in each case, at the longed-for possession of "that special specimen." The use of technical mining and Spanish terms links the specialized knowledge of mineral collecting with the idea of old Mexico and the old mining West.

Also, the reader is depicted as a man of leisure, propping his feet up and reading a magazine, but his leisure taken up not only with the consumption of the products of others' labor, but with the Romantic contemplation of the labor itself in the distance. Another preface in the series equates the miners' labor with the labor of producing the magazine, also consumed by the putative reader:

As the red dusk settles over the quiet desert, you can hear the distant sounds of the miners' pick-axes cutting the limestone gossans in search of pockets, and also the faint tapping sound of us at our computer keyboards documenting it all for yet another special issue. (*Mineralogical Record* 2003a:3)

Clearly, these passages are satirical, in that they intentionally exaggerate an accurate picture of the magazine, capturing many important dimensions of its techniques for distributing social and cultural capital. The capacity for self-satire is also a valued quality for mineral collectors and dealers, who are often willing to appreciate the peculiarities of their avocation, and who are in most cases bemused and interested to find themselves as the "natives" in an anthropological study.[4]

Socializing

Much work to distribute social and cultural capital is done through the social interactions at the various gem and mineral shows and other occasions throughout the year. Although there are many shows in many different cities, and a number of cities have concentrations of high-end mineral collectors (such as Fallbrook, California; Houston, and Denver), I focus here on the Tucson Gem, Mineral, and Fossil Showcase and on activities in Tucson more generally. There are now forty-three gem, mineral, and fossil shows between mid-January and mid-February in Tucson. The gem and mineral show sponsored by the Tucson Gem and Mineral Society was the first of the shows; the other shows are tagging on to the phenomenal success of this show. Over the years, the shows have been starting earlier and earlier, so that the TGMS show culminates what could now be called a mineral season (Villanueva 2007). Among these forty-three shows, there is a broad range, including shows that cater to more casual shoppers, lapidarists, fossil collectors, and many others. Likewise, the forms of social and cultural capital vary. The three top shows, catering to the most elite collectors, are the Westward Look Resort Show, the Tucson Gem and Mineral Show at the Convention Center, and the Arizona Mineral and Fossil Show's location at the Tucson City Center (formerly InnSuites) hotel. For the Westward Look Resort Show, dealers participate by invitation only, and the TGMS show requires that dealers have participated in other shows (a far less stringent requirement). The monetary costs for renting space, equipment, electricity, and such are also about $1,000 higher than at other shows (around $3,000 as opposed to around $2,000) (Ross et al. 2010:171–173). But these obstacles are only a few of those that aspiring dealers have to deal with. More implicit but arguably more influential are the forms of social and cultural capital needed for entry.

In addition to being a marketplace for minerals,[5] the Tucson shows are a constellation of social events and interactions. Although a number of dealers and collectors live in Tucson, they represent only a small fraction of an international collectivity of mineral enthusiasts and professionals, many of whom travel the world to collecting sites and to mineral shows. Though the Internet commerce in minerals has grown tremendously, it does not replace the many occasions for face-to-face contact between mineral lovers. In large part, this is because much of what happens in the mineral business takes place in the context of informal socializing. Indeed, the boundary between socializing and business is almost nonexistent. A brief description adapted from my field notes helps to give a sense of this.

In February 2009, I attended the Tucson Gem and Mineral Show. Upon entering the hall on my first day there, I saw dozens of clots of people chatting away with old friends, catching up on the year's news, and talking about the show and the mineral world in general. I joined a few of these clots myself. Upon arriving at the show my first day, I went to the booth of one dealer I had interviewed the year before. We talked about a contact I had made earlier in the year between a dealer in Mexico who wished to sell her late husband's collection. The material was not as good as promised (he thinks the dealer's "North American wife" [to distinguish her from his "Mexican wife"] got the best minerals), but he thanked me for getting him back in touch with her. He said, "I gave her $500, because you never know." We then discussed the rising incidence of narco-related violence in northern Mexico and his expectations for the show. I told him I was looking to interview more women in the business, and asked if he thought his wife, who shares the business with him, might be willing to be interviewed, and he said he thought she would and then directed me to a couple of other possible informants. I thanked him and bade him good-bye, saying I would circle back later that day to see if she had returned to the booth.

I then went to another booth, where I saw a mineralogist I had interviewed some years before, and a mining engineer from New Mexico, with whom I exchanged business cards. As before, the subject soon turned to violence in northern Mexico. Another dealer, whose name I did not catch, said, "instead of invading Iraq, we should invade Mexico and clean things up there." An awkward silence followed. A few moments later, the mining engineer began talking pointedly, admiringly, about President Obama's speech for Lincoln's birthday that had taken place the night before. The dealer said nothing. I took the opportunity to ask for more details in the convoluted story of the Aztec Sun specimen, and a few minutes later took my leave and headed for another booth. In these encounters, we can see the fluid transitions between business and other topics, in the context of many people congregating who have known each other for years, but who only see each other at these events. (In this respect, the shows resemble academic conferences, also sites for the distribution and management of social and cultural capital.)[6]

Dealers' booths on the Convention Center floor often have comfortable chairs set up so that people can stop and talk. In recent years, some dealers have paired up to rent a larger space and create a seating area together. And in the shows based in hotels, such as the Arizona Mineral and Fossil Show at the Hotel Tucson City Center (formerly known as InnSuites), there is also the chance to hang out in a dealer's room and chat or wait for friends to pass through. Often the dealer will provide soft drinks and sometimes wine or beer to create a sociable atmosphere, which keeps people in the room and near the minerals.

At Tucson and during shows in other places with a concentration of mineral dealers and collectors, there are often parties that take place on different nights of the show. These can become annual events. One party I attended in Tucson was a catered affair with about fifty guests, in the private home of a collector, exploration geologist, and TGMS member. The party brought together dealers, collectors, and museum curators (and—to the best of my knowledge—just one anthropologist). People were dressed in what might be described as cocktail wear with—for instance—jewel-toned evening blouses over nice jeans for women and patterned button-downs and khakis for men. The meal was served buffet style and the guests ate sitting on the couch and chairs or on the floor, cheerfully making room for new diners.

The party could easily have been for lawyers, software professionals, or academics (though this last would entail a sartorial decline). Here, however, much of the conversation concerned mining localities, mineral price trends, and private collections. In one room, a group, mostly men, looked at the host's exceptionally fine collection of Mexican minerals, which consisted of choice, cabinet-size specimens gathered carefully over several decades. In their well-lit cases, each one glowed: a little miracle of color and form. People demonstrated various signs of awe and enthusiasm as they looked at the collection.

The social scene becomes more intense during the TGMS show at the end of the season, but many dealers and serious collectors stay a month or more in Tucson, seeing the shows, catching up with old friends, and enjoying the nice weather. As one Tucsonan told me, "The weather's really great in February, and it's never been an expensive place. They like to talk about how it's gotten more expensive but it hasn't really. It's got friendly people, good restaurants, so people like to spend time here."

Given the nature of the mineral business and its reliance on networks of people and information, the intense sociality of the shows and the events and occasions surrounding them is not the least bit surprising. However, describing a few of these does show the importance of social and cultural capital in these events. The ability to socialize comfortably in middle-class and wealthy circles, to display knowledge not only about minerals themselves, but also about localities, events, and trends in the business, and aesthetic standards all pertain to a class-based habitus far more available to some than others. And class is not the only common ground for participants in this marketplace. Whereas many Europeans attend (particularly from France and Germany), all of those who interact with the high-end mineral world can speak English well. English is a necessary condition for participating in this world.[7] There are occasional uses of particular Spanish phrases or words; these tend either to approach Jane Hill's characterization of "mock Spanish" as a form of (mis)appropriation of Spanish by Anglo Southwesterners (2007)[8] or to be used as a mark of sophistication (but only when accompanied by fluency in English).

It is not, by any means, impossible for a Mexican person to participate in this world; Dr. Miguel Romero (see chapter 4) was a highly respected member of this social world and a good friend to many. But such a person would need to speak English fluently and to possess and display the appropriate habitus; that is, the bearing, deportment, speech, responses, and other embodied forms of middle-class status. One dealer remarked that he had helped another U.S. dealer sell a mineral to the American Museum of Natural History. Asked why the dealer did not sell directly to the museum, he said,

> Well, he was kind of a rough guy . . . he liked his whiskey. I remember
> one time he said something to Paul Desautels [the Smithsonian curator
> of minerals; see chapter 2] here in one of the rooms at Tucson and Paul
> walked out angry, saying "forget this" and that's why the American
> Museum got it [the mineral] instead of the Smithsonian.

This anecdote suggests that the dealer who controlled the minerals may have recognized his own inability to deploy the kind of social and cultural capital necessary to get the prices he wanted. He therefore had his friend (a courtly and engaging person, well used to being around wealthy and influential men) help him out.

Gender

The required habitus for participating in this world includes particular forms of masculinity. This shows itself in a highly gendered language toward the minerals themselves, portrayed either as the female object of male desire or as phallic symbols. For instance, when I asked people why they thought mineral collecting appealed more to men than women, they often used the phrase "mine is bigger than yours" to describe the competitively masculine nature of collecting.[9] The competitive flavor of many interactions among collectors resonates in the United States as particularly male and tends to be less available to—and also sometimes less appealing to—women than to men. Although there are some important women collectors, they are a small minority and many of them see themselves as struggling to gain full access to the world of mineral collecting.[10] Though women attend many of the social events surrounding the show, they do so generally as wives of collectors. At the party I described, there were quite a few men unaccompanied by women, but hardly any women unaccompanied by men. I myself went as a male friend's "date." No one would have cared had I gone alone, but I fit in better at the party with a male companion.

Even when included physically, women are often tacitly excluded by being ignored or discounted. A prominent female collector, who collects along with her husband, has begun publicly pointing out some of the difficulties confronting women collectors, saying that many men (and women) "think that women don't know anything or that they only collect rocks because they're pretty." This impression that women only collect rocks because they are pretty was repeated to me by several people, both men and women. The word *pretty*, used by each of them, captures a sense of feminine triviality. The men who call themselves "aesthetic collectors" also collect rocks because they are pretty, but it is never put in those terms.[11]

To illustrate the way in which women's interest in minerals is ignored, my collector informant told me the following:

> One time a bunch of collectors came over and I cooked and then we talked about the mineral collection and then at the end, the women

said, "your cooking was wonderful and so is your husband's mineral collection." After that, I said to my husband, "We're catering next time. That is not going to happen again."

These various strategies and forms of labor help to monitor and restrict access to the high end of the minerals market at Tucson. They act as gate-keeping practices that allow only certain kinds of people to sell at the high prices in this market. In this way, the opportunity for arbitrage is maintained through the ongoing production of difference based on class, gender, nationality, and race.

In Mapimí, gate-keeping practices take a somewhat different form, focusing on rights to access to minerals. Outsiders can find ways to get access (and producers want to sell as directly as possibly to these outsiders). But producers block access to the mine, and middlemen try to preserve their positions within the supply chain and in doing so increase transaction costs for those from outside. As Christopher Steiner has said, with respect to the African art market, "In the art world, as in all big business, the success of the middleman depends on the separation of buyers from sellers" (1995:151).

Mapimí

Elaborate systems of rights—both formally codified and implicit—regulate access to minerals in Mapimí. These systems rest in part on powerfully held notions of who has legitimate claims over minerals, but also on the ability to durably make those claims through legal channels, control of information, and force (or the threat of force). The whole set of these practices, and the mineral specimens at their heart, are locally known as *"el risco."*

El Risco

The word *risco,* used widely in northern Mexican mining to refer to mineral specimens, loosely translates as *riches* or *treasure* and, like those words, captures a sense of serendipitous fortune. The activity and economy of extracting and selling mineral specimens is also referred to as *el risco.* This usage is comparable to other occupational terms, such as *la mina* (mining) and *el pollo* (working for Trasgo/Tyson chicken). El risco differs from these other pursuits in its inherently sporadic quality, and for this reason it works best if combined with mining, working in tourism, or some other activity. The possibility of hitting the lottery with the discovery of a fantastic legrandite or adamite specimen adds to the dashing quality of el risco. However, even those who have made large sums of money are acutely aware of how much more others, further from the source, have profited from their labor and risk. In our interview with Felix Esquivel, his resentment over the huge sums of money paid for the Aztec Sun was palpable and echoed in a number of other interviews.

One benefit of el risco as an occupation is that the minerals themselves can act as a kind of alternative special-use money. According to one informant, a geologist with decades of experience collecting Mexican specimens, this was especially true in the earlier years of the market:

> Years and years ago when you wanted to find minerals, you went to the bar. The middleman was the guy who owned the bar. The miners had a commodity to offer in exchange that was in very high demand and they could convert specimens to liquor without going through the homefront. That meant they could turn over all their paycheck [to their wives], and the fact that they traded specimens that were six months' pay never impacted the domestic situation.

One miner, recalling his days as a young man working in the cooperative and el risco, said:

> Many times we would sell them [minerals] to buy beer and go out looking for women. We used to go to the red light district in Torreón with the money we made from riscos, and we would always come back broke. If we came back in a taxi, we would have to wake up the moneylenders [in Mapimí] when we got there to ask them for some money. Later we would pay them back with more money we got from el risco or with our wages (from the cooperative).

This appears still to be true, at least to some extent. One mineral dealer in Mapimí said to me, "For them [the miners], it's on top of their salary. Lots of times they sell them so they can go drinking. Sometimes they go out drinking and they bring a box [of minerals] so they can keep drinking."

Here, general-purpose currency intervenes, but the principle of using minerals as a means to go drinking with friends is similar. Minerals thus allow for the expression of certain forms of masculinity that are not oriented to the reproduction of the family, but to male collectivities maintained through drinking and sex (with prostitutes), or what in Mexico is often termed *el desmadre*. El desmadre (literally, "demothered") has multiple valences in Mexico, including mess, chaos, and activities—coded as particularly male—such as drinking, fighting, and what would in the United States be called "raising hell" (Gutmann 2006; Magazine 2007).[12]

In the 1970s and 1980s, the market was more diffuse and many specimens were coming out, including the jackpot strikes of legrandite in 1977 and purple adamite in 1982. Although the prices for these minerals were far lower than they are today, there was greater opportunity for the miners to realize their value. The miner quoted previously remarked, "In those days, el risco was more valuable." The minerals themselves were less costly in real terms, but the relative lack of middlemen made them more profitable for miners such as this one. As we will see, this is not to say that the market at this time was not rigidly controlled at a variety of levels, but that the way in which it was controlled, along with the quantity of good specimens emerging, distributed the profits for miners more widely (though in the more centralized market of today, some miners are doing quite well).

Access to mineral specimens coming out of Ojuela has been managed according to a series of legal and extralegal rights to minerals. Rights to legal access are determined by links to one of several corporate institutions with legal rights to the mine. Once the minerals are taken out of the mine, a complex network of buyers from Mapimí, Bermejillo (the town on the nearest highway, which runs north to Ciudad Juárez), Chihuahua, El Paso, and Tucson regulates access to them. The links between these different groups and institutions are thickly interwoven and often riddled with conflict.

There are also less formal claims to access, based on considerations of moral economy, on daring, and on force. In what follows, I analyze various institutions, norms, and other strategies by which people in Mapimí control access to minerals.

Institutions: Peñoles, Cooperativa, Top-Gem

Industrias Peñoles is the second-largest Mexican-owned mining corporation (the first is Grupo México). It is partially capitalized with Canadian and South African money, but most of the leadership and investors are Mexican. It operated out of Mapimí between 1895 and 1931, building a line from Mapimí to the main railroad at Bermejillo and constructing many of the houses in the eastern end of town. In 1931, Peñoles ceased direct production at Ojuela and moved its headquarters to Torreón. The company grew to be one of the largest Mexican-owned mining corporations and now controls many of the most productive mines in the north of Mexico, including Fresnillo, Zacatecas, which, as of this writing, is the largest silver mine in the world (perhaps soon to be eclipsed by Goldcorp's mine in the same zone, Peñasquita). When Peñoles moved to Torreón, the miners who had been at Ojuela moved down the mountain to Mapimí, mostly to the *colonia mecánica* that had been built by Peñoles during its heyday in the town. The company rented these houses to the miners at a reduced rate, and in the 1990s, it allowed the miners to purchase them.

Between 1931 and 1940, these miners (the ones who did not move away to other cities in Mexico or to the United States or take up other pursuits) eked out a living on the scrap metal left in the mine by Peñoles. By all accounts, things were dismal in those years, and the population of the town shrank by almost 75 percent.

The Sociedad Cooperativa de Producción "Mineros de Ojuela" was founded in 1940 with fifty-seven members. Many of these members were former employees of Peñoles or their sons or nephews, who had moved down from the town of Ojuela to Mapimí after the company moved away. For several decades, Peñoles purchased lead carbonates (cerussite) from the cooperative for use as a flux (lead is a necessary agent in the smelting process) in the plant at Torreón, where all the silver from its northern mines (and mines held by other companies) was processed, which was especially needed because the largest mine, at Fresnillo, produced a silver ore with a very low lead content. This arrangement "ran off the rails," according to one geologist I interviewed, when the company faced heavy fines for the lead and arsenic released in the smelting process.

In addition to selling lead to the foundry, cooperative members could also take out mineral specimens. Some specimens were mined as property of the whole membership and were sold from the old Peñoles social club (*casino*), later used as the cooperative's headquarters. The proceeds from these sales were divided equally among the members or sometimes used to help the families of injured miners. Nevertheless, until the 1970s, mineral prices were for the most part rather low. When North Americans came to Mapimí to buy minerals in the 1950s and 1960s, they even sometimes paid miners in old clothes instead of money. Lázaro de Anda's brother Melquior recalled to me his early life when his father was one of the main mineral dealers in the town,

> There used to be a lot of poverty. People would come to the house of my father [to buy minerals] and give bags of chocolates to us. We learned to say, "one hundred pesos [in English]." They paid only enough [for us] to eat badly, to subsist.

The cooperative folded in 1991, when many of its members were reaching the age of retirement. This transition is a common stress point for cooperatives, and many are unable to make the transition to a new generation of members. One former member described tension between cooperative members and their sons, who were anxious for the older members to retire and leave their places to them.[13]

At this point, a wholesale dealership based in Tucson, called Top-Gem Minerals, Inc., and headed by the Tucson dealer Mike New, moved into town and obtained a lease for specimen mining at Ojuela, as well as permission from the federal government to take the minerals out of the country.[14] New pays his workforce and provides them with social security and in turn they mine specimens for him and sell them to him through the main dealer in town, Lázaro de Anda.

New took charge of the specimen mining in Ojuela, which had been *muck-bound;* that is, mining was increasingly impeded by the presence of the broken rock left in the mine after drilling and dynamiting (known as muck rock). He also took steps to control the flow of specimens out of the mine by paying people in advance to sell the best material to him and by cracking down on risqueros. One miner reported to me:

> Fourteen or sixteen years ago people from outside used to come to the plaza and there would be the risqueros selling riscos. [There were] Germans, French, and gringos, and the *coyotes* [Mexican dealers from outside Mapimí], there were almost always coyotes, because they would go sell at the Tucson show. "OK, gabacho [a slang term for a foreigner, especially from the United States] here are some minerals." The selling on the plaza stopped when . . . a person came who rented the mine with the purpose of taking out riscos. And then the risqueros couldn't go in any more. He monopolized the production.[15]

In 2004, a former cooperative member, newly returned from the United States, formed a partnership with a Monterrey businessman and secured a contract allowing them to mine Ojuela for galena to use in the smelter at Torreón. Silver prices rose by 36 percent between 2003 and 2004, sparking the demand for lead to process silver. In 2007, when I

179

visited Mapimí the first time, there were approximately 150 people working in the mines extracting galena, most of whom were also taking out minerals on the side. In the summer of 2007, a miner could make 3,000–4,000 pesos weekly, compared with 770 per week working for Tyson.[16] In August 2007, silver posted an average price of $12.28 per troy ounce, nearly twice what it was in 2004.

In July 2008, when I returned to Mapimí, silver averaged $18.03, the highest monthly average (not corrected for inflation) since 1980, when the Hunt Brothers tried to corner the market (Silver Institute n.d.). (Silver averaged $35.11 per troy ounce in 2011 and $31.15 in 2012.) Nevertheless, the workforce was down to about eighty to one hundred men, because the high arsenic content in the ore being sent to the smelter incurred high fines there. Peñoles was pressuring the current leasers to install a processing plant at Mapimí (to distance themselves from the environmental problems caused by smelting). Many of the less experienced miners had been laid off and were working for Tyson or migrating to the United States. The remaining miners were also extracting mineral specimens, mostly of higher quality though in less quantity than in former years.

Miners and Risqueros

In 2008, I went back to Mapimí with two research assistants from the Universidad Iberoamericana (Rubén Lechuga and Vera Regehr). We interviewed a dozen miners and dealers and spent part of our time hanging around the casino with the miners. Rubén's presence was especially helpful as it is mostly unheard-of for women to hang around on the street near the casino. I did not feel in any danger when I went to talk to miners at the casino on my previous trip, but my presence was so odd that it constrained conversation. As with the party in Tucson (but even more so) it helped to have a "date" when talking to these informants.[17]

The casino occupies a wedge-shaped lot at the eastern entrance to Mapimí. It has a bronze bust of Pancho Villa in a small terrace in front, raised about three feet above the street. On one side of the casino runs Mapimí's main east-west road and on the other, receding into the brushy desert that surrounds the town, are the remains of Peñoles's offices and smelter. It is here that Pancho Villa dipped his toe into the Peñoles swimming pool in the 1920 photo shown in chapter 1 (figure 1.3).

Miners congregate on the edge of the terrace that fronts the casino and the surrounding curbs. They wait here for the truck that takes them to the mine and hang out and drink beer and soda in the late afternoon once the sun has passed its zenith. In July 2008, when we visited, it was the time of the annual town festival, so there were more people gathered than had been there the previous year. There are also more young men in the town these days. Young men who were working in the United States or elsewhere are returning to Mapimí because of the boom in metals prices and the increased difficulty of migrating (and now because of higher unemployment in the United States).

Here they can watch all the traffic that enters the town, hail their friends, and remark on the timeliness of the bus or the arrival of geologists or other outsiders. They do not seem to miss much. Once Rubén, Vera, and I introduced ourselves, we entered into lively conversation with a group of miners who took care to point out that they had seen me the previous year walking on the highway in the hot sun, wearing a funny hat. But they were also happy to talk to us about their lives and work histories, the trials of being in the United States (though they spoke of this with some constraint), the increased exploration and production in mining, and the trade in riscos.

They also pointed out to us the strongest among them, inviting us to look at their muscles and reporting the weights they could carry (reportedly up to 110 kilos [242 pounds] at a time). Several noted the danger of trying to carry too much; one said that those who push themselves too hard "break down their body and pretty soon by doctors' orders they can't work in the mine anymore." Nevertheless, it is clear that masculinity, strength, and physical courage form part of the necessary habitus for those who wish to get access to minerals through their role as miners.

On this trip, I was able to go down into the mine. As we descended in the skip, the mine foreman, who had graciously agreed to take us around, told us of a previous visit from an engineer from Peñoles: "He couldn't believe it. He said it looked like a book he had of how the mines used to be historically, in the nineteenth century." Indeed, the conditions in Ojuela are extraordinarily rudimentary, especially when compared with the period from 1895 to 1931, when Peñoles operated the mine directly. At that time, there was electricity installed in the mines, and mining cars running along an underground rail system. Now there is no electricity, and much of the rail was sold as scrap after Peñoles's departure. The new company has installed a mechanized skip, but the miners use hammers and picks to knock out the rock and to set the dynamite, and their headlamps are powered by carbide, a system that went out of use in most mines in the Americas in the early twentieth century.[18] There are also mules stabled down in the mines to help tow some of the cars.

The mine is enormous, with huge caverns and snaking passageways, and it is easy to see both that it is very old and that for decades it has only been exploited sporadically. It is also clear that there are countless entrances to the mine (and these can be seen from outside as well) through which risqueros might enter. On one wall is nailed a mummified bat; the foreman tells us that it was already there when he first began working in the mine as a young man (he is now in his fifties). Here and there are mangled beams, bits of machinery, and lengths of iron rail. One iron barrel is stamped with a swastika; it apparently dates from the late 1920s, during which time Peñoles was partly capitalized by German investors.

We go down to the water level, which is lower than it has been for years because of the ongoing drought in the region. (This is one reason why better mineral specimens are starting to come out in the last few years, as areas formerly submerged become accessible.) We can see the old pumping machinery, unused since the 1930s (figure 6.2). Its massive, High Modern forms remind me that this was once a mine that produced on a large scale and justified intensive capital investment. The black water lapping at the iron beams covers more levels, more tunnels, and more riches below.

The work that these miners do is extraordinary. The skip works primarily to move personnel, and much of the labor of actually carrying the lead out of the mine is done by men, carrying it in sacks by tumpline (a sack with a strap that hangs from the top of the carrier's head, a technique inherited from Pre-Columbian times). One miner told me proudly, ruefully, that he has to climb thirty-six ladders from his work area to the surface, carrying about one hundred kilograms by tumpline. The total distance from his work area to the surface (vertical and horizontal together) is about a kilometer. When he said this the evening I met him, I was impressed, but nothing like as shocked as when I saw the men at work. There were obviously enormously strong and yet as they carried the sacks, sweat poured off them and they staggered, resting frequently and groaning aloud with the weight. We were not surprised when our

FIGURE 6.2. Water level, Ojuela Mine, Mapimí, Durango, Mexico.
Photo by Elizabeth Ferry.

guide told us that one of the most common injuries was to the kidneys, as a result of this tremendous weight. Later, several men told us that they usually made five to six trips per day, carrying a total daily of over five hundred kilos. However, in 2007 and 2008, they could make more in two to three days at this than in a full week working for Tyson.

Besides the job of carrying out the ore (*cargador*), there are also those who break it from the rock and set the explosives (*tumbadores*) and their assistants. We visited one tumbador at his work area, accessed via a series of ladders and bridges into a small, steaming cavern. Don Pepe, the former president of the cooperative, was celebrating his eighty-second birthday on the day we visited. Diminutive but powerfully built, he continued to break rock and set charges two days a week, even though the foreman and mine superintendent had offered him the chance to retire and "rest in his house" at full salary.[19] He was too hard-of-hearing to carry on much of a conversation in the mine itself, with all the noise of the work around us, but in a later interview, he told me that he simply preferred to keep working in the mine.[20]

Along with the pay from the production of galena, these miners also take out mineral specimens, selling them to Lázaro, to other Mapimí dealers, or to one of the coyotes who come down from Juárez or Chihuahua. Each work team (tumbador, ayudante, and cargadores) has rights to the minerals found in their work area, and the profits are supposed to be divided equally among the team. This system is similar to the one operating in Guanajuato (where the perforista and ayudante have rights to the minerals) and in other Mexican mines.

Those who are officially engaged in mining ore are not the only ones extracting specimens.[21] As in the past, along with the various players embedded within the institutional structures of mining Ojuela, there are also risqueros. Risqueros do not have formal claims on the mine, but rather they have the daring, knowledge, and social networks to enter the mine at other entrances or at other times to extract specimens. These men were somewhat impeded by the activities of Top-Gem in the 1980s and 1990s, but they have flourished again since the suspension of organized specimen mining. In some cases, what they lack in official access to minerals they can make up for in skill at mining specimens, which is a far more delicate process than producing ore. For instance, one man, who has been working in el risco for over thirty years, criticized the miners working for Mike New and for the contracting company, saying, "with all due respect, they don't know how to take out mineral specimens."

Many risqueros live in the other part of Mapimí, to the west of the plaza, and do not have historical ties to the cooperative. The town is in some ways divided between these groups, with the former cooperative members and current miners having more economic and political power, but the risqueros continuing to work and sometimes grabbing a prize out from the miners' noses, such as the Aztec Sun.

Miners and risqueros, then, draw on different sources of social capital in their claims over mineral specimens. Many of the current miners have some connection to the cooperative and to Peñoles (by descent, because many of their fathers and grandfathers worked for the company, and by residence, because many live in the neighborhood built by the company). Their claims over mineral specimens draw on these formal connections to Ojuela, as well as the guildlike sense of collectivity that characterizes many mining communities. Not all sons of miners have gone into mining; many prefer to seek work in the United States or to work in the chicken business. But their familial history is nonetheless intimately connected to the industrial history of the mine and the town, and this history (and their current employment as miners) gives them access to minerals.

Risqueros gain access to minerals by entering at night or between shifts through one of the multiple adits let into the mountainside. They have no formal claim on the mine and do not benefit from the social standing that the cooperative still seems to hold in the town. However, there are also risquero families, so the social capital that risqueros draw is also partly kin-based, and there seems to be a fund of knowledge and a way of knowing the mine that is available only to risqueros.

I asked one risquero if it was difficult to extract riscos, and he replied: "It's all difficult—you have to know how to cut [the rock] and take out [the riscos]. You have to learn. . . . Older people taught me—my uncle and my cousins." Later, in describing how risqueros enter the mine, he said, "The mine is like a city. The different entrances are like different neighborhoods." The first statement here emphasizes the importance of knowledge for risqueros, both of the work itself and of the mine. This knowledge is only partially available to those who work the mine in more conventional ways, as miners for the company or cooperative. His description of the mine as a city further underscores his sense of himself as an urbane resident of the mine. He went on to describe nights he had slept in the mine, places where he left clothing to change into, and so on. This knowledge is not formalized by some external authority. It is passed down through the family and hard-won through experience, and it entails a wiliness that is not necessary for miners.

Comerciantes and Coyotes

Miners and risqueros alike sell to Lázaro and to other dealers in the town, to U.S. dealers, and to the coyotes. These coyotes also purchase directly from Lázaro; according to him, they tend not to pay as well as the U.S. buyers. He said:

> The Americans always pay me and the workers [miners] well. The ones who don't pay well are the ones we call "coyotes." They come from Juárez, from Chihuahua, from Zacatecas and they want to buy low and sell very high, and that's where the problem is.

The tension between dealers at different parts of the commodity chain runs through all of these kinds of markets. Mineral prices are so changeable and so dependent on complex questions of access and knowledge that the number of middlemen in a given commodity chain can make a huge difference in prices and in the distribution of profits. Competition between dealers within Mapimí, from elsewhere in Mexico, and from the United States (particularly El Paso and Tucson) has characterized the mineral business since its emergence in the 1940s. In the early years, the market was largely controlled by dealers in El Paso and Tucson who imported minerals from northern Mexico, including Mapimí, and sold them in Tucson.

In fact, as described in chapter 1, the flow of Mexican minerals helped to get the Tucson show off the ground in the 1950s. El Paso and Tucson dealers quickly became some of the most influential in moving minerals to Tucson. Many of them also employed Mexican agents to help find localities and make contacts. One of the most famous of these is Shorty Bonilla, based in Chihuahua. George Griffiths in Gómez Palacio also helped to build the mineral market in Mapimí by purchasing minerals and sometimes bringing North American dealers and collectors to Mapimí to acquire specimens directly.

The dealer Rock Currier reports on these early years in Mapimí, saying, "Specimens [from Ojuela] were sold in *rejas,* a sort of large standardized wooden box of the kind normally used for shipping produce. Thousands of rejas of adamite, hemimorphite, aurichalcite, calcite, etc., known as the 'Mapimí mix' were produced." This description accords with the sense of profusion of minerals in these years and with their low prices relative to those of later years (after the legrandite and purple adamite finds) (Currier 2008:28).

Sometimes violence or the threat of violence (physical or otherwise) helped a dealer keep outsiders from gaining access to cheap minerals. A family in Bermejillo also competed to control the local mineral business. One member of this family, described as a *cacique* (chief or boss) by several, kept a strong hold on much of the trade in the years before Top-Gem's arrival, at times resorting to threatening physical violence. The dealer Colonel Baron, according to one informant, "really had a lock on El Paso. If he found out about other people bringing in minerals he would have his friends on the Mexican side throw them in jail."

As everyone knows, the U.S.-Mexican border does not stop extra-legal movements of goods and people, but channels those movements and captures the profits from them in particular ways. The legality of exporting minerals is somewhat foggy, because they are technically part of the subsoil, and therefore national patrimony, but in practice they have rarely been treated as such. However, they are subject to customs inspections and possible impounding at the border, and U.S. mineral dealers who wish to bring them across often have Mexican brokers in Juárez to help them. These people are known as *brincadores* (border hoppers). Brincadores possess particular kinds of social and cultural capital, including speaking fluent Spanish and a characteristically male, northern Mexican habitus that allows them to maneuver through legal obstacles more smoothly (and probably more inexpensively, because they are not charged the "gringo tax").

However, there were always U.S. dealers coming down. In addition to Gene Schlepp and later Mike New, there were John Whitmire, Bill Panczner, Jack Amsbury, Curt Van Scriver, and others who were all over northern Mexico buying minerals. When U.S. (Anglo) dealers bring minerals across the border themselves, they must draw on their own knowledge of the border and how it works.

Though mineral commerce between Mapimí and Tucson has waxed and waned, with phases of greater or less monopolization, there has been all along a thicket of both Mexican and U.S. dealers and their agents jockeying to gain access to minerals at low prices while keeping others from doing the same, through legal structures, social networks, and force and the threat of force.

There is some indication this situation may be changing. The specimen market in Mapimí declined between the mid-1990s and mid-2000s, though several recent finds have given it new life. As one informant put it,

> specimen markets have become worse for Ojuela and Guanajuato [my other main Mexican field site], because supply has dwindled at the big mines. Buyers are taking fewer trips and the trips are longer and less profitable. Partly this is because of the Internet. And at this point a lot of the middlemen have been to Tucson.

This suggests that ongoing arbitrage between the markets may be dissolving the price gap, which is consistent with the predictions of economic theory. But this pressure toward equalization is constantly constrained by the careful arrangement and deployment of social and cultural capital. No one person can buy (or extract) at the extreme low end of the market and sell at the extreme high end, though they can make a good deal of profit by inserting themselves closer to these endpoints. They are thus still practicing arbitrage but under conditions of constraint. Even when people "jump the gate," which they sometimes do, they are often subsumed into supplier relationships or forced to pay transaction costs. How? Through the ongoing production of class, gender, racial, and national difference.

Jumping the Gate

Since the 1980s, Mike New of Top-Gem Minerals has carried out a concerted campaign to gain and maintain control of the supply of minerals at Mapimí. He has done this by penetrating each step of the process, obtaining a license to mine for specimens, paying his miners a retainer and a share in the profits, and helping to support the main dealer in Mapimí, Lázaro de Anda. He also operates a wholesale dealership in Tucson, which imports not only from Mapimí, but from all over the world. According to a number of informants in Mapimí, Top-Gem's activities have fundamentally changed the market in Mapimí, making it more strictly regulated and centralized and displacing earlier channels for minerals controlled by local dealers. In this endeavor, he encountered a lot of resistance, especially from a dealer in Bermejillo who controlled much of the trade in earlier decades. This dealer combated New's entrance into the market through legal measures based on claims that the minerals, as national patrimony, could not leave the country, or that New could not hold the license he holds because he is not a Mexican citizen, through placing articles in the paper criticizing New, and through threats of force. In an interview in February 2009, New described a showdown with this dealer:

> He came at me with a machete. I turned around and ran to my car—he thought I was running away, but I came back with a .357 Magnum. . . . Then he brought a lawsuit against me. We stopped working and I hired an attorney. . . . I talked with the officials of Mapimí. I met with the risqueros.

Mike New's case shows what happens when a North American dealer infiltrates the spaces normally reserved for miners and risqueros, the challenges he faced, and the strategies he used to overcome those challenges.

However, he has not been able to access these minerals without cost. In an online post reporting on his interview with Mike New in 2010, Eric Greene of Treasure Mountain Mining enumerates the various challenges of moving minerals from Mexican mining localities to the United States. He concludes by saying,

> Amazingly, Mike talks about dealing with these potentially deadly challenges in a matter of fact, no nonsense way that contrasts sharply with the fear I had imagined lurked at every turn in the process. Of course, he has been traveling to Mexico to buy minerals for as long as anyone can remember, and has developed the savvy and street smarts to survive in these challenging conditions. I'm perfectly content to let him do the mineral shopping in Mexico, and then I buy my favorites for a reasonable wholesale price, and *no* risk. (Greene 2010)

Mike New has been able to get his hands on minerals closer to the source than many other U.S. dealers. But he does this through paying a certain number of transaction costs that are not paid by those in Mapimí. He can still arbitrage the minerals by selling them in Tucson, but his ability to buy cheaply in Mapimí is constrained; this also helps to keep the price gap open. In addition, as a primarily wholesale dealer, he does not typically sell directly to those buyers willing to pay the highest prices, though he has also sold some of the finest Mapimí specimens ever (such as the purple adamites found in 1981) and these were certainly not sold at wholesale prices.

Such infiltrations happen in the other direction as well. Pepe Villegas, a miner from Milpillas, Sonora (a new copper mineral locality producing fabulous malachite and azurite pieces), has been selling his specimens directly at Tucson and other shows. Milpillas is an exciting new locality because it is part of the province of porphyry copper deposits that spans Arizona and has yielded classic malachite and azurite from Bisbee and other places (Moore and Origlieri 2008). In the 2000s, when minerals began coming out, there was a miniature rush to the area. However, the barriers to access are high. Dealers from both Mexico and the United States had been attacked on the road from Milpillas, beaten, and robbed of their specimens. Around this time, Villegas drove north with his specimens to the 2009 Tucson show. He met with the show promoter of the Arizona Mineral and Fossil Show (held every year at the InnSuites hotel), who let him set up a cheap booth next to the hotel entrance. This angered some of the dealers inside, particularly those who also were selling copper minerals from Milpillas, but who had to pay more for their space and whose specimens would only be seen later, after buyers entered the hotel.

I spoke with Villegas and his daughter outside the Arizona Mineral and Fossil Show, one of the most successful of the hotel-based shows. He described the conflicts over minerals at Milpillas, and then asked me about Mapimí, with an eye to going there to buy minerals. He said, "There are always conflicts in this business, but there's always a way to do it if one wants to." I also heard from another source that he was buying minerals at Fresnillo. Like Mike New, he is trying to establish more direct channels between different markets, but he is trying to keep those channels to himself. However, with limited English and little familiarity with the culture of mineral shows in the United States, he is unable to command the right social and cultural capital. His habitus conforms to a class, racial, and national position that tends to work against his ability to charge high prices. One dealer put his finger on this when he went to the show promoter and, as an oblique complaint about the choice to give Pepe Villegas a place at the entrance, put on a Rasta hat with fake dreadlocks and said, "Hey mon, I got some cool rocks, put me in front, mon," much to the promoter's annoyance. The promoter continued to rent Pepe a place outside the entrance in future shows, but this incident shows that at least some in Tucson considered him to be out of place.

Villegas continued to work other shows, but with limited success. One dealer reported in an e-mail:

> The market started to saturate. The major players had lots of malachite and not enough azurites. The lots from [Pepe] who now controlled most of the action were big bucks and it was harder to sell. Then the market crashed. Pepe, finding no buyers started selling out of his truck and then the hotel entrance and he sold stuff for in some cases less than wholesale. Wholesale buyers are pissed. At Costa Mesa [the West Coast Mineral and Fossil Show in Costa Mesa, California] he is on second floor with higher prices. On Sunday he packs up and goes home. Tells staff his boy has to be in school Monday. Jungle drums say he sold one rock. . . . The mine will have a short specimen production life and the good ones will be future classics.

The strategy of charging less than wholesale (thus preventing other middlemen from entering the commodity chain) angered other dealers and made them unwilling to help Villegas learn the ropes of the Tucson market, and this may also have contributed to some of the troubles he had maintaining prices. However, I wonder whether this fact, along with market saturation, completely accounts for the fall in Milpillas prices, or if the fact that a Mexican miner was widely seen selling them made people less willing to pay the highest prices. As one informant told me, diagnosing the reason why Mexican minerals are rather inexpensive in comparison with minerals from other parts of the world, "Mexico is our poor neighbor to the South, so everything from there is supposed to be cheap."[22] By displaying the cultural capital of a market close to the mine, rather than that of Tucson or another international show, Villegas may have inadvertently triggered the kinds of gate-keeping practices I described earlier, and for that reason, contributed to the decline in prices.

As of this writing (2011), the Milpillas prices have recovered. This is in part because of new finds and the release of minerals that had formerly been held back from the market. Also, several other dealers, including Evan Jones, have managed to get hold of rich sources for Milpillas minerals. Indeed, Villegas may now be one of Jones's main suppliers, though he also continues to sell on his own. I am speculating here, but it may be that the market is more robust in part because he is now selling more of his stock through an Anglo dealer who does control the necessary social and cultural capital (indeed, Evan Jones is the son of Bob Jones, the mineral collector and author who gave the Aztec Sun its name).[23] If this is so, once Villegas is back in his "rightful place" in the background, the minerals themselves seem able to shed their history, as minerals tend to do.[24]

The gate-keeping practices in each market continue to be effective enough to maintain price differences even when the boundaries between them are breached. Thus it is not that conventional economic theory about arbitrage is wrong, but that other factors constantly work against the degradation of opportunities for arbitrage. These include the deployment of social and cultural capital linked to distinct categories of race, class, nation, and gender. Such categories, then, are not "externalities," but they are integral to the production and maintenance of the two markets.[25] And not only that, but the practices that make and link the two markets also reinforce these forms of difference.

Conclusion

In Mapimí, people need access to specimens, and thus the regulation of social and cultural capital centers on rights and claims to access. The forms of social and cultural capital at issue include kin and occupational networks, knowledge of Mexican law, membership in the cooperative, knowledge of the mine, and certain forms of working-class northern Mexican habitus. In Tucson, to enter into the marketplace, one must have as a prior condition access to either minerals or money. But not everyone who has minerals can play. Here the relevant social and cultural capital takes the form of cultivated ideas of taste and connoisseurship and an Anglo middle-class habitus, including fluency in English.

In a recent article, "Beyond Economic and Ecological Standardization: Supply Chains and the Human Condition," Anna Tsing argues for the importance of difference in supply chains and the contemporary form of capitalism of which they form such a large part. She writes:

> thinking through supply chains offers an opening to reconsider the relationship between culture and economy. Supply chains *depend* on those very factors banished from the economic; this is what makes them profitable. Supply chains draw upon and vitalize class niches and investment strategies formed through the vicissitudes of gender, race, ethnicity, nationality, religion, sexuality, age and citizenship status. We cannot ignore these so-called "cultural" factors in considering the mobilization of labor. (2009:165)

The stories in this chapter demonstrate that the maintenance of a price gap between Mapimí and Tucson and the persistence of opportunities for arbitrage also depend on such "cultural" factors. From mine through marketplaces, people create value for minerals by exploiting difference, not only between prices but also between people. The transactions between the markets help to create and maintain the salience of these differences.

We can think of the different sites along the mineral commodity chain that runs from Mapimí and Tucson as nodes where labor to create value converges in especially dense clusters. At sites such as Mapimí, El Paso, and Tucson, people and objects create value in ways that are perforce different than in other sites. This is because making such distinctions is vital to the gate-keeping practices that make arbitrage possible. For instance, connoisseurs in Tucson seek to preserve their own privileged capacity to judge pristine beauty in minerals, whereas miners and risqueros in Mapimí assert their official status, knowledge of the mine, and male courage and strength as giving them a distinct claim over minerals. They ratify meaningful distinctions between minerals, between markets, and between people. In doing so, they reproduce those distinctions, which come to seem like rock-hard facts or like the bounded entities we used to call *cultures*. This seeming rock-hardness and boundedness is an effect of successful value-making, but because value-making is inherently unstable and emergent, it is by no means unchanging.

CONCLUSION

I conclude with two images of Mexican minerals: one moving through space and the other through time.

Two Stories

In the spring of 1998, as we were preparing to return to the United States after twenty months of fieldwork, my husband and I gave a party at our house to say good-bye to our neighbors and friends. We rented chairs and a tent, hired a band, and engaged our neighbor and friend Paco to kill a pig and make *carnitas*. Our guests included our neighbors in the town of Santa Rosa de Lima, miners, and other cooperative members, faculty and students from the University of Guanajuato, and a *regidor* (alderman).

Although it is not as common for people to bring gifts to a party (such as a bottle of wine or flowers) as it is in some U.S. contexts, one cooperative member, who worked in the automotive department, brought me a small rock wrapped in tissue paper. It was a specimen of native silver growing out of a base of black acanthite (silver sulfide). The silver looked like the slightly curved bristles on a toothbrush. The specimen came from the El Cubo mine, he told me, and he wanted to give it to me to remind me of my friends in Guanajuato. I was delighted to receive this gift and carried it back to the United States with pride.

For several years, the mineral played a role in my interviews with dealers and collectors; I would show it to them during the interview and it often provoked conversation. They were especially interested to hear that it came from El Cubo, which is not known for its silver and acanthite specimens. I kept it on my bureau with my jewelry, and it traveled to five different states as my husband and I chased postdoctoral fellowships and visiting professorships around the country.

Eventually settling in Boston, and beginning to include museum collections in my study of minerals, value, and transnational space, I got to know Dr. Carl Francis, the curator of Harvard University's mineral collection. I interviewed him formally several times, audited his course on collecting and curating, and consulted him on many questions related to this project. I found his view on museum collecting both sensible and inspiring, and I noted the traces of Guanajuato in the collections with interest (see chapter 4).

After thinking it over for some months, I decided to donate my specimen to the Harvard collection. I was no longer in touch with the person who had given me the stone, but I decided that by donating the piece I would be linking it to other small chunks of Guanajuato that had fetched up in the Harvard collections over the past 130 years. This seemed like something that the giver would not disapprove of. By placing the piece in the company of other Guanajuato minerals, I felt I was giving it a good home.

When I showed the mineral to Carl, he said,

> I really like this specimen, because it tells a story. Not every specimen does tell a story. This is acanthite [pointing to the part on the bottom], a silver sulfide. Where it was the surrounding environment was so low in sulfur that it sucked the sulfur out of the specimen, leaving silver wire.

This mineral actually tells at least two stories. One concerns the interaction of elements over enormously long periods of time. The mineral seems like the quintessence of unchanging matter, but it is actually a production of prior forces in an agonistic encounter now stabilized in its current form. Carl's description calls attention to the story of the interaction of silver, sulfur, and the surrounding environment that resulted in this specimen.

The other story takes place in a far shorter time frame and concerns the city of Guanajuato and its mines as a site of mining over several hundred years and as the site of anthropology fieldwork in the late twentieth century. The mineral traveled to the United States because of both the mining history and my fieldwork, and it arrived at Harvard because of other minerals that had journeyed from Guanajuato to Cambridge. The deed of gift that I signed to transfer legal ownership of the specimen records a brief version of both stories:

> One mineral specimen received as a gift from Miguel Arreguín in Santa Rosa, Guanajuato, Mexico, during fieldwork in June 1998.
> Silver wires extruding from acanthite from El Cubo mine at Guanajuato, Guanajuato, Mexico (Deed of gift from Harvard University to Elizabeth Ferry and David Wood).

The two stories are connected. Silver brought mining to Guanajuato, obviously. Mining made the mineralogist Severo Navia start a collection that he later sold to Albert Holden, who bequeathed it to Harvard. The dominance of U.S. mining in the early twentieth century brought Frank DeVotie, a banker, to the Guanajuato mining district (Herr and Herr 1999:152), where he sold a specimen to Harvard mineralogist Harry Berman; that specimen is still in the Harvard collection. The existence of the silver mining Santa Fe Cooperative brought me to Guanajuato in 1996. The distinctiveness of my silver-acanthite specimen made it a good going-away present.

In short, the presence and interaction of sulfur and silver made possible U.S.-Mexican transnational connections that have left traces in the form of the collections, acquisition records, my interviews, and this book. Each of these traces results from the making of a different form of value: silver as commodity, a thoughtful going-away gift, the Harvard mineral collections, an ethnography, and so on. These forms of value came together in the mineral's removal from the mine, its travels, and its eventual coming to rest at Harvard. In its small way, this mineral's journey linked Guanajuato and Cambridge as places within transnational space.

Social-Material Worlds

One of the caves at the Naica mine in Chihuahua, where enormous gypsum crystals have formed under conditions of high heat and humidity (see chapter 3), is known as La Cueva de las Velas, the Cave of the Sails. It is named for what speleologists Tullio Bernabei, Paolo Forti, and Roberto Villasuso have described as "a completely new type of gypsum speleothem [a secondary mineral formation within a cave]: the 'sails'" (Bernabei et al. 2007:23). Their article describes the formation of extremely delicate (less than 1-mm-thick) sail-like structures on top of some of the gypsum crystals, those whose terminations are directed upward (figure C.1).

The sails were formed by evaporation, which induced "an upward air current controlled by convective motion of moist air" (p. 28). Capillary action helped to draw gypsum crystals upward, forming the sails. This could only have taken place when the water level of the mine decreased so as to create an aerate environment, which happened in the 1980s when the mine was pumped out to a level below the cave. The authors note two further things about the sails: first, that "at least presently, the Cueva de

FIGURE C.1. "Sails" gypsum formation, la Cueva de las Velas, Naica, Chihuahua, Mexico. Photo by Giovanni Badino / La Venta Esplorazioni Geografiche. Reprinted with permission.

las Velas is the single cave, not only within the Naica mine but in the entire world, in which these peculiar speleothems have had the unique possibility to develop" (p. 29); and second, that they are doomed. As the cave is ventilated, it is cooling, and when the temperature falls below the dew point, "strong condensation will occur with the consequence of a fast and complete dissolution of the sails" (p. 29). The sails' uniqueness, their luminous fragility, and their oncoming death give this plainspoken speleological description an unexpected poignancy.

The sails are manifestly the product of multiple human and non-human actors and forces, including gypsum, water, air, heat, capillary action, evaporation, the price of silver in the 1980s (which impelled Industrias Peñoles to drain the mine), and the activities of miners, mining engineers, and speleologists (who need the cave to be ventilated and therefore cooled in order to do their work). These actors and forces have convened in response to prior acts of value-making that established silver as a global commodity and speleology as a branch of science that merits research funding and publication. All these things together have co-alesced in the sails, delightfully surprising formations that have a life and tell a story. The sails are meaningfully distinct from and also intimately connected to their environs. In their tiny orbit, they are social-material worlds that have and make value.

APPENDIX:
SOURCES AND METHODS

Data on Guanajuato come from fieldwork in the summers of 2001, 2003, and 2007, as well as from twenty months of dissertation fieldwork focused on the Santa Fe Cooperative (1996–1998) on which my previous book is based. Data on Mapimí, Durango, are drawn from two visits in the summers of 2007 and 2008 and from interviews in Tucson with U.S. and Mexican dealers who work in or travel to Mapimí.

Archival research in the summer of 2005 forms the source for the Smithsonian material. In addition, I attended the Tucson Gem and Mineral Show (2004, 2009), the Denver Gem and Mineral show (2005), and the East Coast Gem and Mineral Show (2003, 2004, 2008), and participated as a member (and newsletter editor) of the Boston Mineral Club for two years (2005–2006). I also conducted archival research at the British Museum of Natural History, the American Museum of Natural History (in New York City), the Harvard Mineralogical Museum, the Palacio de Minería, the Archivo Histórico Genaro Estrada in Mexico City, and the Museum of Mineralogy "Eduardo Villaseñor Söhle" at the Guanajuato School of Mines.

These diverse sources have naturally called for a range of research methods, of which the most important has been semistructured interviews. In Mexico and the United States, I have conducted approximately sixty interviews with miners, dealers, and collectors, as well as more informal conversations and e-mail exchanges. The sessions have ranged from twenty minutes to over two hours. The majority of the interviews were audio-recorded with supplementary written notes (on some occasions, ambient noise or technical difficulties prevented audio-recording or the interviewee asked not to be recorded). Of these subjects, about ten people have become more established "informants" with whom I have had several to dozens of conversations (and to whom I owe particular thanks).

Participant observation for this project has included: entering the Valenciana and Rayas mines in Guanajuato and the Ojuela mine in Mapimí to observe the mining process and mineral specimen extraction; observing dozens of sales and other interactions between merchants and buyers in Guanajuato, Mapimí, and at the various gem and mineral shows;

visiting miners in their homes to see altars and displays with minerals, and collectors in their homes to see collections; socializing with miners, dealers, and collectors on a variety of occasions, both in Mexico and the United States; attending meetings and presentations of the Boston Mineral Club and at the shows; editing the BMC newsletter; and taking a course in collecting and curating at the Harvard University Extension School, taught by Dr. Carl Francis, the curator of the Harvard Mineralogical Museum.

The excellent journal *Mineralogical Record* has been a rich source of historical information as well as contemporary ethnographic data on mineral collecting in the late twentieth and early twenty-first centuries. The videos produced by the company BlueCap Productions have also provided much valuable information. I have also regularly perused up to twenty websites of mineral dealers, messages on the rockhounds.com listserv and the online mineral collectors' community Mindat.org.

NOTES

Introduction

1. While this definition is still used by the U.S. Bureau of Mines (which publishes the *Dictionary of Mining, Mineral, and Related Terms*), more recently the definition has been broadened to include some organically formed minerals, minerals that are amorphous (rather than crystalline) in structure, and "liquid crystals." The broader definition currently accepted is: "A mineral is an element or compound, amorphous or crystalline, formed through biogeochemical processes" (Skinner 2005). However, because the instances covered in this expansion lie far outside the realm that is the focus of this book, I have chosen the older, serviceable though not completely exhaustive, definition.

2. A productive third line of enquiry in twentieth-century anthropology has focused on inalienability, locating value in the restrictions placed on exchange and in the dialectical relationship of "giving" and "keeping" (Ferry 2005b; Godelier 1999; Mauss 1990 [1950]; Weiner 1992). Some formulations within this tradition suggest a possible rapprochement of the debate between value as rooted in labor and in desire/exchange (e.g., Godelier 1999). However, this discussion lies beyond the scope of my introduction here.

3. The reader may notice that my description here partly resembles the work of the structural linguist Ferdinand de Saussure on "linguistic value," defined as the position of a given phoneme or word within a system of oppositions. Saussure writes, "Language is a system of interdependent terms in which the value of each term results solely from the simultaneous presence of the others, and in particular from its distinction from those other terms" (Saussure 1959 [1916]:121). At the same time, only some distinctions can be said to have value, because only some are recognized as meaningful within a given linguistic system.

4. Callon et al. (2002) presume no essential division between those who bring a good to market and those who choose between different potential goods in the qualification process. This feature resonates well with the role of connoisseurship in the contemporary mineral market, as we will see in chapters 2, 5, and 6.

5. Actor-network theory has been criticized for an overly economistic view of actors and their commitments, essentially treating both human and nonhuman actors as self-maximizing individuals (Martin 1995). However, if we see that the emergence of stable convenings of actors into network occurs within a world created by earlier such convenings—including such things as nation-states, trade agreements, ideas of racial and national fitness, and so on—we see that actors cannot exist as independent self-maximizing gamesters but are constantly embedded within prior formations (themselves actors, too) that both motivate and limit their actions.

1. Histories, Mineralogies, Economies

1. This book focuses especially on the extraction and movements of ore minerals, because these are associated with larger-scale mining efforts. The fact that they are considered by some to be the waste or by-product of the mining process, and that they come to have so many different forms of value, makes them especially useful for a study of value-in-formation.

2. The appendix contains a detailed description of my sources and methods.

3. cuentame.inegi.org.mx/monografias/informacion/gto/poblacion/default.aspx?tema=me&e=11.

4. Azurite = $Cu_3[CO_3]_2(OH)_2$; malachite = $Cu_2(CO_3)(OH)_2$; cuprite = Cu_2O.

5. My informants did not all agree that Mexican minerals played a formative role in the development of the show. One very knowledgeable person pointed out that the presence of Mexican minerals was not often cited as a reason for the show's popularity (as are, for instance, the climate and the role of Paul Desautels). However, he did go on to say that it could be a contributing factor, even if not recognized as such by participants.

6. http://mapserver.inegi.gob.mx/AHL/realizaBusquedaurl.do?cvegeo=100130001.

7. In July 2007, workers made 3,000–4,000 pesos per week, as compared to 770 pesos made at the Tyson farms. In July 2008, the pay is roughly the same, although the mine has had to cut the workforce from about 140 to 100 because of excessive arsenic in the ore, which results in penalties by the foundry.

8. This is because of the wide range of specimens extracted in a find, and the need of sellers to move inventory, as well as to establish a steady income alongside the possibility of a great strike.

9. The descendants of these two sources are the National Museum of Natural History and the National Museum of American History (formerly the Museum of History and Technology). A proposed National Museum of Engineering and Industry was never established (Molella 1991).

10. In fact, James Smithson's work focused on mineralogy and chemistry. His private mineral cabinet and catalogs came to the Smithsonian along with the bequest, but these were destroyed in a fire in the Smithsonian building in 1865. In 1832, the mineral *smithsonite,* a zinc carbonate, was named for him.

11. For a detailed discussion of the ways minerals have been exhibited at the Smithsonian and the significance for understandings of nature, see Ferry 2010.

2. Shifting Stones

1. Neptunism was the ascendant theory in the late eighteenth century, but it was challenged by Volcanists who posited that many rocks (particularly basalt) were formed from cooled lava, and Plutonists, most prominently James Hutton, who placed particular emphasis on the inner heat of the earth. These theories were superseded by Charles Lyell with the publication of *The Principles of Geology* in 1830 (Guntau 1987).

2. Due to a series of errors and mischances, Del Río never received official credit for his most famous mineralogical discovery. In 1801, he isolated a new element in the lead deposits at Zimapán, Hidalgo. Humboldt, however, upon examining the samples, tentatively identified them as impure chromium, with no evidence of a new material. Del Río revised his earlier claims and described the substances as a "subchromate" of lead. In 1830, Nils Gabriel Sefström of Sweden rediscovered the element Del Río had called erytronium in a Swedish deposit, naming it vanadium after Vanadis, the Scandinavian goddess of beauty. Upon analysis, this turned out to be the same as the mineral Del Río had found at Zimapán. Several attempts through the years to change the name back to erytronium and give credit for discovery to Del Río did not, unfortunately, succeed (Bargalló 1966:44–45).

3. The Dana system classifies minerals in terms of their negative ions (anions), based on the discovery that minerals with similar anions also tend to have similar properties.

The positively charged ions (cations—mostly metals) indicate the particular species within the larger class of minerals. For example, within the sulfide class (in which the anion is sulfur), we find different mineral species such as acanthite (silver sulfide), galena (lead sulfide), pyrite (iron sulfide), and pyrargyrite (silver antimony sulfide).

4. The names Instituto Geológico de México and Instituto de Geología de México appear to have been used interchangeably.

5. Thanks to Engineer Oscar Irazaba for much of the information in this section (personal communication).

6. http://www.mnh.si.edu/onehundredyears/expeditions/Paricutin.html.

7. In the letter, Foshag also mentions that González Reyna was trained at the Guanajuato School of Mines, where he must have studied with Ponciano Aguilar.

8. Much more could be and indeed has already been said about the nuances and implications of these shifts, but for our purposes, a broad-brush description is enough. The citations in this section form an introductory guide to further reading for the interested reader.

9. This value of exemplarity clearly plays in the various forms of scientific observation and depiction described by Daston and Galison in their study *Objectivity* (2007).

10. I have found very little evidence of *Mexican* mineral dealers prior to the 1940s (see chapter 6).

11. The National Museum of Natural History was reorganized in 1963, and the Department of Geology was split into two entities: the departments of Paleobiology and Mineral Sciences, of which the Division of Mineralogy forms part.

12. Desautels acted as a private agent for Perkins Sams during his tenure as Smithsonian curator. This activity, as well as some other purported infringements of Smithsonian policy and ethical codes—de-accessioning gems from the Isaac Leys collection in contradiction of the terms of the bequest, and participating in inflated appraisals of minerals donated to the museum (for tax benefits to the donor)—led to his forced resignation in 1983. He died in 1991.

13. In his article "Connoisseurship in Minerals," Wilson remarks that "most connoisseurs . . . are confident and capable of making the necessary fine distinctions; where their standards for repair are fully met, they feel no need to penalize the specimen" (Wilson 1990:10).

3. Making Scientific Value

1. Indeed, in the catalogues of his collection, numerous specimens from San Carlos are recorded, with what appear to be prices that range from 0.5 to 25 pesos (presumably he paid these amounts to miners in exchange for the specimens, because there is no evidence that he ever sold minerals; if so, this is the earliest evidence I have found of a mineral specimen market in Guanajuato).

2. Thus, acanthite/argentite, one of the more common silver minerals found in Guanajuato, is silver sulfide; pyrargyrite is silver antimony sulfide; and proustite is silver arsenic sulfide.

3. The recently described species miguelromeroite honors the Mexican collector Miguel Romero (see chapter 4).

4. Lorraine Daston describes an "anxiety over galloping synonymy [more than one name for the same species] among nineteenth-century botanists, leading to the development of international codes for botanical description and nomenclature" (2004:154). The CNMNMN is part of the same effort, but some time afterward. Pete

J. Dunn (who has perhaps the largest number of new mineral descriptions to his credit of anyone alive) gives a thorough account of the trials that must be undergone by an "unknown" mineral in order to receive recognition and a new name (Dunn 1977).

5. Aguilar became a prominent member of the political and social elite of Guanajuato. He was a member of the social club Casino de Guanajuato, president of the group Pro-Guanajuato (the first organization in the city dedicated to urban preservation), and vice president, in 1923, of the Concurso Fraterno de los Hijos del Colegio del Estado. His position in the city was sufficiently notable that in 1914 he was jailed by the revolutionary government of the city along with seven other prominent figures. He remained in jail for ninety-seven days (Jáuregui 2002:47–49).

6. Many thanks to James Holstein, collections manager of Meteoritics and Mineralogy of the Field Museum, for his generous help tracking down the type specimen.

7. The full text of the treaty can be found at http://www.mexica.net/guadhida.php.

8. The map located the Rio Grande at El Paso as two degrees west and El Paso itself as thirty miles north of their actual locations (Kohout n.d.).

9. For example, Bartlett (1965 [1854]), Emory (1987 [1857]), Hewitt (1990), Jiménez (1857), Rebert (2004), Salazar Ylarregui (1851), Tamayo P. de Ham and Moncada Maya (2001), and Werne (1987).

10. Because this study concerns rocks and minerals, I focus on the attempts to create geological and mineralogical value. However, one could tell the story with reference to zoology, ethnology, topography, or botany with different results.

11. It does seem that there was a geologist assigned to the survey at the beginning. His name was Theodore Moss and he left or was dismissed in 1851. From Santa Rita del Cobre, New Mexico, on August 14, 1851, Colonel Graham, head of the scientific corps attached to the survey, wrote to ask Bartlett to allow Moss to take back the "minerals and fossils he has collected while attached to the commission." Bartlett responded, rather tartly, that Moss would have to apply to him directly for permission (this is clearly part of a larger conflict between Bartlett and Graham that resulted in Graham being replaced on the commission by William H. Emory) (U.S. Army Corps of Topographical Engineers and Graham 1799–1865). I have found no further record of those collections.

12. Salazar's Conservative associations resulted in his appointment by Emperor Maximilian as Imperial Commissioner for Yucatán from 1864 to 1866. Following the defeat of Maximilian in 1867, Salazar was exiled to New York for a time but returned to teach until his death in 1892 (Arrigunga 1996).

13. Berlandier spent the last years of his life in Matamoros, Mexico, where he amassed a personal collection of minerals and other specimens of natural history. Darius Couch, who purchased the collection and donated most of it to the Smithsonian, wrote to Spencer Baird saying, "I purchase it because it's too valuable to general science to let remain and rot in Matamoros" (SIA Record Unit 7002, box 18, folder 8, letter from Darius Couch to Spencer Baird, February 15, 1853). This remark reiterates the idea of Mexico as unsuitable for the production of "general" (universal) scientific value.

4. Mineral Collections and Their Minerals

1. The Roebling collection (along with that of Frederick Canfield, fortuitously bequeathed in the same year, 1926) catapulted the Smithsonian into the forefront of world collections.

2. This is not quite an accurate portrayal of Aguilar's legacy, for two reasons. First,

Aguilar gave a lecture on this technique to the Instituto Geológico de México on September 20, 1919 (CPA, box 59), eleven years earlier than the letter reports; and second, however prescient his work may have been, "men in later years," at least those outside of Mexico, did not know they were following his lead, nor is Aguilar recognized in the genealogy of atomic absorption spectrophotometry.

3. This letter and the others in this section are back-translations, because the original letters in English are not in the archive, but only the translations made by Glenn for Aguilar.

4. The catalogue given to the university at that time, prepared by María Aguilar Zavaleta, reported the following contents of the collection:

Catalogued specimens: 6,210

Of these, 416 specimen numbers include 2, 3, 5, 12, and more examples, giving a total of 1,020 additional specimens.

Specimens without numbers listed in the catalogue and found on various shelves: 92

Specimens without numbers, not already mentioned: 73

Total: 7,395

In addition a French collection of rocks and minerals: 300 specimens

French collection of fossils: 200 examples. (CPA, box 95)

5. There are now over four thousand recognized mineral species.

6. The report goes on to acknowledge that because the field of mineralogy has moved away from descriptive mineralogy, the collections are less central to the scientific work of the university than they used to be. However, the rise of amateur collecting and the emergence of minerals as a valuable asset class (see chapter 2) have also opened up new opportunities for exhibition and education. The report recommended that the museum be curated in the future by a person entirely dedicated to that endeavor (and thus not a member of the research faculty), and that the priorities of the museum be reorganized, away from research and toward public education. The university accepted these recommendations for the most part. Since 1998, Harvard Museum of Natural History handles the exhibition and educational aspects of the natural history collections (including the Museum of Comparative Zoology, the Harvard University Herbaria, and the Mineralogical and Geological Museum), but the curatorial functions remain separate.

7. The 2006 dispersal of the Academy of Natural Sciences of Philadelphia collections provides a counterexample. This collection, founded in 1812, encompassed the collections of many early U.S. collectors, including Adam Seybert, the first person in the United States to be trained formally as a mineralogist (he was trained in Europe after receiving a medical degree from the University of Pennsylvania in 1794); A. E. Foote, a prominent nineteenth-century mineral specimen dealer; William Vaux, a wealthy Philadelphia collector; and Joel Roberts Poinsett, minister to Mexico from 1825 to 1830. The Philadelphia Academy Collection also incorporated the earlier American Philosophical Society collection, which went back to pre-Revolutionary times. In 2006, the collection was sold because there were not enough funds to curate and steward the pieces; it was bought by a consortium of three mineral dealers: Collector's Edge, Kristalle, and Crystal Classics. An article posted on Crystal Classics' website details the collection's history and the process of dispersal, stating:

What we hoped to do was to take this collection and restore it as best we could, by cleaning, repairing, and relabeling with today's mineral names and locations. The aim was to pass several suites to museums and institutions to be curated and displayed on view for the public once again; and also to bring parts of the collection to the market, so that mineral collectors would have the chance to own a piece of this significant history. And by preserving the history with each specimen, we can in part preserve the whole collection even though it is no longer together. (Crystal Classics 2007)

This final claim to "preserve the whole collection" through the specimen characterizes what often happens when a famous collection is broken up.

8. Dealers also likely wanted the distinction of having their specimens in the Romero collection, as the ability to place specimens in famous collections is itself of great value.

9. This sense of collections as spheres of exchange underlies the ethical codes of many museums and collections.

10. For detailed discussions of gender and mineral collecting, see chapter 6 and Ferry 2011.

5. Making Places in Space

1. On a more scholarly plane, world-systems theory and its many analogues and offshoots echo this view. The vocabulary of *periphery* and *core* or *metropole* and *hinterland* is premised on theories about the movement of value from some places in the world (often in what we now call the Global South) to others (in the Global North) through the conversion of raw materials into finished commodities.

2. In this respect, my view of value-making as action differs from that of David Graeber, who places intention squarely at the center of his theory of value, as part of his own utopian political project (2001). I do not disagree with that project, but it is not quite the same as my own.

3. In the mine at Elmwood, Tennessee (a zinc mine that also produces fine specimens of fluorite, barite, sphalerite, and calcite minerals prized by collectors), the mine owners set up an employee profit-sharing plan: The mine managers sold all of the specimens and the proceeds were divided at the end of the year into equal shares for all employees; one share for each worker (management was not included). The penalty for violating the rules by the theft of specimens by miners was instant dismissal (Lawrence H. Conklin, personal communication).

4. Now Santa Rosa is also home to workers in the Cerámica de Santa Rosa, as well as those who work in Guanajuato City and León. It is also the parish seat for some twenty mountain villages in the Sierra de Santa Rosa and a weekend getaway for prosperous Guanajuatenses. See Ferry 2005b for a full description.

5. To show how gifts can hold people at arm's length even as they bind them together, David Graeber uses an example of the custom among middle-class Americans of bringing a wine bottle to dinner, which marks a new phase of life when they no longer freely wander in and out of each others' houses (2001:226–227). Minerals in Guanajuato do something similar by establishing men and women as being on friendly terms, but neither as lovers nor relatives.

6. I use Collector's Edge as the example of this strategy, as they are acknowledged to have pioneered it in the 1990s, but now most high-end dealers present their miner-

als in this way. Increasingly, dealers band together to rent floor space so as to create more spacious and inviting showrooms.

7. The mounts are designed by the Sunnywood Collection, a company that specializes in the custom design of wooden and acrylic mounts for high-end specimens and bills itself as "specialists in presentation of natural art" (sunnywood.com/default.asp).

8. Christopher Steiner has noted a similar process in the alteration of objects for the African art market to make them seem more "natural" and "authentic" (Steiner 1995).

9. Chapter 6 discusses the *Mineralogical Record* in more detail.

10. Warwick Anderson's discussion of the circulation of blood and tissue samples of Kuru sufferers among U.S. biologists as a "gift-economy" provides a compelling analogy to the ways collectors circulate minerals (2008:133–134).

11. Moreover, given that minerals are extracted through mining and that many collectors have some ties to the mining industry, it seems quite telling that those most responsible for blasting, grinding up, and melting down the earth are also most enchanted by the pristine qualities of minerals.

6. Mineral Marketplaces, Arbitrage, and the Production of Difference

1. Note the description of offshore production as based on "geographical" difference, as if a disparity in labor markets were a natural feature of the landscape.

2. Some interviewees made it a condition of being interviewed that they would not answer questions about specific prices, and my informants in general did not want me to publish that information.

3. For more information on the awards won, please refer to Allen Press's website (allenpress.com/ news/07272009).

4. When I told one prominent collector about the Kula ring, remarking on its similarity to the circulation of high-end mineral specimens, he found the connection both fascinating and amusing (in much the same way as I did) and carefully noted the bibliographic reference.

5. Many other things are also sold in these shows, most importantly fossils and gems. These objects lie outside the scope of this book. See Villanueva 2007 for an account of the full range of activities that take place at the Tucson shows.

6. Participant observation is especially easy in these events, for a primary activity of an anthropologist in the field, tramping about talking to people, is also the activity of the other people at the show.

7. In this discussion, as in much of the rest of the book, I am referring only to one segment of the mineral business, as described in chapter 2. The trade in minerals is multitiered and takes many different forms. I am primarily interested in understanding what allows people to participate at the top of the market.

8. Hill describes the ways in which this kind of speech contributes to a broader system of white superiority, even when used "by people who are not, personally, racist in any of the usual senses" (Hill 2007:272).

9. For a full discussion of gender and mineral collecting see Ferry 2011.

10. This has not gone unnoticed. There is even a DVD produced by BlueCap Productions (which makes DVDs for the mineral world) called *Perspectives of the Woman Collector*.

11. This distinction is similar to that drawn between seamstresses and tailors in nineteenth-century France, as described by Joan Scott (1989).

12. The structuring of multiple gendered currencies in the formation of collectivities also recalls a number of other sociocultural contexts, such as the male and female valuables exchanged in the Trobriand Islands (Weiner 1988) and the circulation of moonshine and cash in rural Russia (Rogers 2005).

13. This conflict seems to contrast with the issues at play in the silver mining cooperative at Guanajuato. Fathers tended to cede place to their sons, nephews, and godsons fairly easily, but they also attempted to get these relatives easier and safer jobs on the surface, so that the ratio between surface workers and underground miners became untenable over time. Based on earlier research (Ferry 2005b), I would argue that the difference lies in part in a more robust idiom of patrimony in Guanajuato, which would help cooperative members see their possessions as belonging to a diachronic corporate group spanning different generations. However, further research into the idioms of patrimony in Mapimí would be necessary to confirm this idea.

14. Top-Gem's workforce is also organized as a cooperative, though it has never had more than ten members. It is not completely clear whether this is, in legal terms, a new cooperative or a new iteration of the old cooperative. There are two conflicting opinions about this, which I have not been able to resolve; New states that it is the same cooperative, whereas the miners insist that the cooperative closed in 1991. All of the members of the new cooperative, as far as I know, were in the former cooperative. However, many of the former members are not now in the cooperative, which, moreover, functions very differently from the old cooperative, because it is capitalized by and sells exclusively to New.

15. For a time, New had a partner from Ciudad Juárez, Rubén Avila. But several years ago he discovered that Rubén was keeping the best specimens for himself, and they parted company. (The formation and breaking up of partnerships like this is a very common occurrence in mineral dealing). In September 2008, Top-Gem sent the following message to its online subscribers:

> To our valued clientele
> In June of this year, Top Gem Minerals, Inc. severed its long relationship
> with Ruben Avila [no accents in original] of Juarez, Mexico.
> Top Gem Minerals, Inc. will not be responsible for any actions by Mr. Avila.
> You will do business with him at your own risk. (e-mail message from Top-
> Gem Minerals to their subscription list, September 26, 2008)

16. In the summer of 2008, the company paid 1.50 pesos per kilogram to break the rock in the mine (*la tumbada*) and 1.50 pesos per kilogram to bring it to the surface (*la sacada*). Miners worked in teams of three to four, who could produce three to four metric tons (three thousand to four thousand kilos) of material per week, for a weekly take of 9,000 to 12,000 pesos, divided equally among the team members.

17. In this case, it was especially helpful to have my "date" be an accomplished anthropologist.

18. Later in the tour, just to keep things interesting, I accidentally lit my research assistant's jeans (her favorite pair) on fire.

19. I am not sure how this would be calculated, because this is a piecework system.

20. I have heard that Don Pepe has since died.

21. In Guanajuato, people do enter the mines secretly, but they do so mostly to extract ore, not mineral specimens. Mineral specimens from Guanajuato are not as expensive as those from Mapimí, which probably accounts for the difference.

22. The assumption that things from Mexico "are supposed to be cheap" resembles the devaluation of Mexican-ness in the U.S. Southwest that makes Zapotec weavings a cheaper alternative to Navajo textiles in U.S. markets (Wood 2008:98–99).

23. Not all of those in the role of suppliers to the high-end dealers are Mexican. Most smaller dealers (who also may have less access to the cultural and social capital characteristics of the southwestern upper middle-class that I described) also sell some of their inventory to higher-end dealers.

24. Elsewhere I have written about the temporalities of mineral specimens, especially their ability to extend their lives as they travel from collection to collection (Ferry 2008).

25. The debate between Garrett Hardin and Elinor Ostrom (and others) over the commons provides a good analogy to my argument. Garrett Hardin described "the tragedy of the commons" (1968), or the idea that individuals using common resources will inevitably degrade or destroy those resources through acting in their own self-interest, even though it is in the interest of all to maintain those resources. He concludes that only privatization or governmental control will adequately protect resources. In a recent article (just to name one of Ostrom's many works on the subject), she and Xavier Basurto do not deny this proposition, but they do point out that many other factors can mitigate or work against it, and that many groups can solve the problem of overuse of resources through "a diversity of norms and rules" (Basurto and Ostrom 2009:255).

REFERENCES

Albuquerque Museum
  1996   Drawing the Borderline: Artist-Explorers of the US–New Mexico Boundary Survey. Albuquerque: University of New Mexico Press.
Anderson, Warwick
  2002   Introduction: Postcolonial Technoscience. Special issue, "Postcolonial Technoscience." Social Studies of Science 32(5/6):643–658.
  2008   The Collectors of Lost Souls: Turning Kuru Scientists into Whitemen. Baltimore: Johns Hopkins University Press.
Appadurai, Arjun, ed.
  1986   The Social Life of Things: Commodities in Cultural Perspective. New York: Cambridge University Press.
Arizona Mining Association
  2006   A History of Mining in Arizona. http://www.azmining.com/images/HISTORY_FULL.pdf, accessed July 10, 2010.
Arnaíz y Freg, Arturo
  1936   Andrés Manuel del Río: Estudio biográfico. Mexico City: Casino Español de México.
  1948   Don Andrés del Río, descubrimiento del Eritronio (Vanadio). Mexico City: Cultura.
Arrigunga, Maritza
  1996   The José Salazar Ilarregui Papers. The Compass Rose 10(1). http://libraries.uta.edu/speccoll/crose96/ilarregui.htm, accessed August 8, 2011.
Asbury Park Sunday Press
  1969   Giant Splashes of Mineral Coloring Embellish Fall Fabrics. July 6.
Azuela, Luz
  2009   La Geología en México en el Siglo XIX: Entre las Aplicaciones Prácticas y la Investigación Básica. Revista Geológica de América Central 41:99–110.
Bakewell, Peter J.
  1984   Miners of the Red Mountain: Indian Labor in Potosí, 1545–1650. Albuquerque: University of New Mexico Press.
Bargalló, Modesto
  1966   Andrés Manuel Del Río y su Obra Científica. Mexico City: Cía. Fundidora de Fierro y Acero de México.
Barrow, Mark V.
  2000   The Specimen Dealer: Entrepreneurial Natural History in America's Gilded Age. Journal of the History of Biology 33:493–534.
Bartlett, John Russell
  1965 [1854]   Personal Narrative of Explorations and Incidents in Texas, New Mexico, California, Sonora, and Chihuahua, Connected with the United States and Mexican Boundary Commission during the Years 1850, '51, '52, and '53. Chicago: Rio Grande Press.
Basch, Linda, Nina Glick Schiller, and Cristina Blanc-Szanton
  1994   Nations Unbound: Transnational Projects, Postcolonial Predicaments, and Deterritorialized Nation-States. New York: Gordon and Breach.

Basurto, Xavier, and Elinor Ostrom
2009 The Core Challenges of Moving beyond Garrett Hardin. Journal of Natural Resources Policy Research 1:3255–3259.

Baudrillard, Jean
1994 The System of Collecting. *In* The Cultures of Collecting. John Elsner and Roger Cardinal, eds. Pp. 7–24. Cambridge: Harvard University Press.

Bennett, Tony
1995 The Birth of the Museum: History, Theory, Politics. New York: Routledge.

Bernabei, T., P. Forti, and Roberto Villasuso
2007 Sails: A New Gypsum Speleothem from Naica, Chihuahua, Mexico. International Journal of Speleology 36(1):23–30.

Bernstein, Marvin D.
1965 The Mexican Mining Industry, 1890–1950. Albany: State University of New York Press.

Bideaux, Richard A., and Terry C. Wallace
1997 Arizona Copper. Rocks & Minerals 72(1):10–27.

Blanco, Mónica, Alma Parra, and Ethelia Ruiz Medrano
2000 Breve Historia de Guanajuato. Mexico City: Fondo de Cultura Económica.

Bohannan, Paul
1955 Some Principles of Exchange and Investment among the Tiv. American Anthropologist 57:60–70.

Bourdieu, Pierre
1986 The Forms of Capital. *In* Handbook of Theory of Research for the Sociology of Education. J. E. Richardson, ed. Pp. 241–258. New York: Greenwood Press.

Brading, David
1971 Miners and Merchants in Bourbon Mexico, 1763–1810. Cambridge: Cambridge University Press.

Brandes, Stanley
2006 Skulls to the Living, Bread to the Dead. New York: Wiley-Blackwell Press.

Brown, Gary
2005 Catspaw Minerals, comment on listserv, April 15, http://rockhounds.ning.com, accessed August 1, 2010.

Callon, Michel, Cécile Méadel, and Vololona Rabeharisoa
2002 The Economy of Qualities. Economy and Society 31(2):194–217.

Cancian, Frank
1994 The Decline of Community in Zinacantan: Economy, Public Life, and Social Stratification, 1960–1987. Palo Alto: Stanford University Press.

Casey, Edward
1996 How To Get from Space to Place in a Fairly Short Stretch of Time: Phenomenological Prolegomena. *In* Senses of Place. Steven Feld and Keith H. Basso, eds. Pp 13–52. Santa Fe, NM: School of American Research Press.

Castañeda, Alejandra
2006 The Politics of Citizenship of Mexican Migrants. New York: LFB Scholarly Publishing.

Chibnik, Michael
2003 Crafting Tradition: The Making and Marketing of Oaxacan Wood Carvings. Austin: University of Texas Press.

Colección Ingeniero Ponciano Aguilar Frías
    N.d.   Archivo de la Dirección de Archivos y Fondos Históricos de la Universidad
    de Guanajuato. Guanajuato, Mexico.
Collins, Jane
    2003   Threads: Gender, Labor and Power in the Global Apparel Industry. Chicago:
    University of Chicago Press.
Crang, Philip, Claire Dwyer, and Peter Jackson, eds.
    2004   Transnational Spaces. New York: Routledge Press.
Crystal Classics
    2007   Philadelphia Academy of Natural Sciences Mineral Collection, August 31.
    http://crystalclassics.co.uk/news-story.php?id=27, accessed February 23, 2009.
Curators' Annual Reports
    N.d.   United States National Museum, Smithsonian Institution Archives Record
    Unit 158. Washington, DC.
Currier, Rock
    2008   Mapimí Legrandite and the "Aztec Sun." *In* The Miguel Romero Collection
    of Mexican Minerals. Terry C. Wallace, Wendell E. Wilson, Peter K. M. Megaw,
    and Rock H. Currier, eds. P. 28. Tucson: Mineralogical Record.
Dana, James Dwight
    1837   System of Mineralogy. New Haven, CT: Durrie and Peck and Haven and
    Noyes.
Daston, Lorraine
    2004   Type Specimens and Scientific Memory. Critical Inquiry 31:153–182.
Daston, Lorraine, and Katharine Park
    1998   Wonders and the Order of Nature, 1150–1750. New York: Zone Books.
Daston, Lorraine, and Peter Galison
    2007   Objectivity. Cambridge, MA: Zone Books.
Dear, Michael
    2005   Monuments, Manifest Destiny and Mexico. Prologue Magazine 37(2).
    http://archives.gov/publications/prologue/2005/summer/mexico-1.html, accessed
    February 12, 2011.
De Cserna, Zoltan
    1990   La Evolución de la Geología en México (1500–1929). Revista del Instituto de
    Geología 9(1):1–20.
De Genova, Nicholas
    2005   Working the Boundaries: Race, Space, and "Illegality" in Mexican Chicago.
    Durham: Duke University Press.
Degoutin, N.
    1912   Les grottes a cristaux de gypse de Naica. Revista de la Sociedad Científica
    Antonio Alzate 32:35–38.
Desautels, Paul E.
    1972   The Mineral Kingdom. New York: Madison Square
Desautels, Paul E., and Roy S. Clarke
    1963   Re-Examination of Legrandite. American Mineralogist 48:1258–1265.
Díaz-Barriga, Miguel
    2008   Distracción: Notes on Cultural Citizenship, Visual Ethnography, and Mexi-
    can Migration to Pennsylvania. Visual Anthropology Review.

Dilworth, Leah, ed.

    2003   Acts of Possession: Collecting in America. New Brunswick, NJ: Rutgers University Press.

Dunn, Pete J.

    1977   From Unknown to Known: The Characterization of New Mineral Species. Mineralogical Record 8:341–349.

Durnett, Raymond E.

    2009   Mining Tycoons in the Age of Empire: Entrepreneurship, High Finance, Politics and Territorial Expansion. London: Ashgate Publishing.

Earley, J.

    1950   Description and Synthesis of the Selenide Minerals. The American Mineralogist 35:360–362.

Eiss, Paul K., and David Pedersen

    2002   Introduction: Values of Value. Cultural Anthropology 17(3):283–290.

El Correo

    2009   Museo de Mineralogía, de los mejores de AL, July 13.

Eliade, Mircea

    1962   The Forge and the Crucible: The Origins and Structure of Alchemy. Chicago: University of Chicago Press.

Elsner, John, and Roger Cardinal, eds.

    1994   The Cultures of Collecting. Cambridge: Harvard University Press.

Emory, William H.

    1987 [1857]   Report on the United States and Mexican Boundary Survey. Austin: Texas State Historical Commission [facsimile edition].

Evans, Anthony

    1997   An Introduction to Economic Geology and Its Environmental Impact. New York: Wiley-Blackwell Press.

Ferry, Elizabeth Emma

    2005a   Geologies of Power: Value Transformations of Mineral Specimens from Guanajuato, Mexico. American Ethnologist 32(3):420–436.

    2005b   Not Ours Alone: Patrimony, Value and Collectivity in Contemporary Mexico. New York: Columbia University Press.

    2008   Rocks of Ages: Temporal Trajectories of Mexican Minerals. In Timely Assets: The Politics of Resources and Their Temporalities. Ferry and Limbert, eds. Pp. 51–74. Santa Fe: SAR Press Advanced Seminar Series.

    2010   Like Ziegfeld Girls Coming Down a Runway: Exhibiting Minerals at the Smithsonian. Journal of Material Culture 15(1):30–63.

    2011   Waste and Potency: Making Men with Minerals in Guanajuato and Tucson. Comparative Studies in Society and History 53(4):914–944.

Findlen, Paula

    1995   Possessing Nature: Museums, Collections and Scientific Culture in Early Modern Italy. Berkeley: University of California Press.

Finn, Janet

    1997   Tracing the Veins: Of Copper, Culture, and Community from Butte to Chuquicamata. Berkeley: University of California Press.

FMR Associates

    2007   Economic Impact of the Gem and Mineral Show. Report for the Metropolitan Tucson Convention and Visitor's Bureau.

Fortun, Kim
    2009   Scaling and Visualizing Multi-Sited Ethnography. *In* Multi-Sited Ethnography: Theory, Praxis and Locality in Contemporary Research. Mark-Anthony Falzon, ed. Pp. 73–86. London: Ashgate Publishing.

Foshag, William F.
    1927   The Selenite Caves of Naica, Mexico. American Mineralogist 12:232–252.
    1928   The Minerals of Mexico. Explorations and Field-Work of the Smithsonian Institution in 1928. Washington, DC: Smithsonian Institution Press.

Foshag, William F., and Jenaro González Reyna
    1956   Birth and Development of Paricutín Volcano. USGS Bulletin 965-D. Pp. 355–489.

Foster, Robert J.
    2007   The Work of the New Economy: Consumers, Brands and Value Creation. Cultural Anthropology 22(4):707–731.

Francis, Carl, Terry Wallace, Peter K. M. Megaw, and Michele Hall-Wallace
    1999   Silver Mineralogy of Guanajuato Mining District, Guanajuato, Mexico. Special issue, "20th Annual FM-TGMS-MSA Mineralogical Symposium: Minerals of Mexico," Mineralogical Record 30(2):84–85.

Franco Ibarra, Jesus
    1997   Potencial Minero de Guanajuato. Guanajuato: Dirección de Minas, Secretaria de Desarrollo Económico.

Franklin, Sarah
    2007   Dolly Mixtures: The Remaking of Genealogy. Durham: Duke University Press.

Frondel, Clifford
    1988   The Geological Sciences at Harvard University from 1788 to 1850. Earth Sciences History 7:1–22.

Gait, Robert
    1999   People for Whom Mexican Minerals Have Been Named. Rocks & Minerals 74(1):44–50.

García-Ruiz, Juan Manuel, Roberto Villasuso, Carlos Ayora, Angels Canals, and Fermín Otálora
    2007   Formation of Natural Gypsum Megacrystals in Naica, Mexico. Geology 35(4):327–330.

Genth, F. A.
    1891   Aguilarite, a New Species. American Journal of Science 141:401–403.

Gereffi, Gary, and Miguel Koerzenewiecz, eds.
    1994   Commodity Chains and Global Capitalism. Westport, CT: Praeger Publishers.

Ghemawat, Pankaj
    2003   The Forgotten Strategy. Harvard Business Review, November:77–84.

Godelier, Maurice
    1999   The Enigma of the Gift. Chicago: University of Chicago Press.

Goldring, Luin
    2000   The Gender and Geography of Citizenship in Mexico-US Transnational Spaces. Identities 1(1):1–37.

Graeber, David
    2001   Toward an Anthropology of Value: The False Coin of Our Own Dreams. New York: Palgrave Press.

Granovetter, Mark

   1985   Economic Action and Social Structure: The Problem of Embeddedness. American Journal of Sociology 91(3):481–510.

Gray, Julian

   2002 [1999]   Finding the Right Mineralogy Text: Dana's System of Mineralogy. The Georgia Mineral Society. http://gamineral.org/Dana-system.htm, accessed August 8, 2011.

Great Panther Silver Limited

   N.d.   http://www.greatpanther.com/s/Guanajuato_Mine.asp?ReportID=411944, accessed August 1, 2011.

Greenblatt, Stephen

   1991   Resonance and Wonder. *In* Exhibiting Cultures: The Poetics and Politics of Museum Display. Ivan Karp and Steven D. Lavine, eds. Pp. 42–56. Washington, DC: Smithsonian Institution Press.

Greene, Eric

   2010   Buying Minerals in Mexico. http://blog.treasuremountainmining.com/2010/11/19/buying-minerals-in-mexico/, accessed June 4, 2011.

Greene, John C., and John G. Burke

   1978   The Science of Minerals in the Age of Jefferson. Transactions of the American Philosophical Society 68(4):1–113.

Gregory, C. A.

   1997   Savage Money: The Anthropology and Politics of Commodity Exchange. Amsterdam: Harwood Academic Publishers.

Guevara Sanginés, María

   2001   Guanajuato diverso: Sabores y sinsabores de su ser mestizo (siglos xvi a xvii). Guanajuato: Instituto Estatal de la Cultura de Guanajuato.

Guntau, Martin

   1996   The Natural History of the Earth. *In* Cultures of Natural History. N. Jardine, J. A. Secord, and E. C. Spary, eds. Pp. 211–229. Cambridge: Cambridge University Press.

Gupta, Akhil

   1992   The Song of the Non-Aligned World: Transnational Identities and the Re-Inscription of Space in Late Capitalism. Cultural Anthropology 7(1):63–79.

Gutmann, Matthew

   2006   The Meanings of Macho. Being a Man in Mexico City. Berkeley: University of California Press.

Hall, James

   1987 [1857]   Geology and Paleontology of the Boundary. *In* Report on the United States and Mexican Boundary Survey. William H. Emory, ed. Pp. 101–140. Austin: Texas State Historical Commission [facsimile edition].

Hannerz, Ulf

   1996   Transnational Connections: Culture, People, Places. New York: Routledge Press.

Hardin, Garrett

   1968   The Tragedy of the Commons. Science 162(385):1243–1248.

Harvard University Geological and Mineralogical Museum

   N.d.   Harvard University Geological and Mineralogical Museum archives, Cambridge, MA.

Hayden, Corinne P.
2003   When Nature Goes Public: The Making and Unmaking of Bioprospecting
in Mexico City. Princeton: Princeton University Press.
Hayward, M. W., and W. H. Triplett
1931   Occurrence of Lead-Zinc Ores in Dolomitic Limestones in Northern Mexico. AIME Technical Publications 442. Englewood, CO: The American Institute
of Mining, Metallurgical, and Petroleum Engineers.
Helmreich, Stefan
2009   Alien Ocean: Anthropological Voyages in Microbial Seas. Berkeley: University of California Press.
Helms, Mary
1988   Ulysses' Sail: An Ethnographic Odyssey of Power, Knowledge and Geographical Distance. Princeton: Princeton University Press.
Henson, Pamela M.
1995   "Objects of Curious Research": The History of Science and Technology at
the Smithsonian. Isis 90:249–269.
2004   A National Science and a National Museum. *In* Museums and Other Institutions of Natural History, Past, Present and Future: A Symposium Held on the Occasion of the 150th Anniversary of the California Academy of Sciences, June 16–17,
2003. Alan Leviton and Michele L. Aldrich, eds. Pp. 34–58. San Francisco, CA:
California Academy of Sciences.
Herr, Robert Woodmansee, and Richard Herr
1999   An American Family in the Mexican Revolution. Lanham, MD: Rowman
and Littlefield Press.
Hewitt, Harry P.
1990   The Mexican Boundary Survey Team: Pedro García Conde in California.
Western Historical Quarterly 21:171–196.
Hill, Jane
2007   Mock Spanish: A Site for the Indexical Reproduction of Racism in
American English. *In* Race, Ethnicity, and Gender: Selected Readings.
Joseph F. Healey and Eileen O'Brien, eds. Pp. 270–285. New York: Pine Forge
Press.
Hirsch, Jennifer
2003   A Courtship after Marriage: Sexuality and Love in Mexican Transnational
Families. Berkeley: University of California Press.
Hoke, George E.
2006   The Railroads of the Compañía Minera de Peñoles. http://mexlist.com
/penoles/index.htm, accessed May 12, 2011.
Hondagneu-Sótelo, Pierrette
1994   Gendered Transitions: Mexican Experiences of Immigration. Berkeley:
University of California Press.
Huginnie, Andrea Y.
1991   "Strikitos": Race, Class, and Work in the Arizona Copper Industry, 1870–
1920. Ph.D. dissertation, Department of Political Science, Yale University.
Impey, Oliver, and MacGregor, Arthur
1985   The Origins of Museums. Oxford: Oxford University Press.
Ingold, Tim
1993   The Temporality of the Landscape. World Archaeology 25(2):27–174.

Irazaba-Ávila, Oscar, and Luis Espinosa Arrubarrena
2002 El museo del Instituto de Geología los gabinetes de mineralogía y paleontología, en Maravillas y Curiosidades mundos inéditos de la Universidad Antiguo Colegio de San Idelfonso, Dirección general del Patrimonio Universitario, UNAM. Pp. 89–108.

Jáuregui, Aurora
1985 Evolución de la Cooperativa Minera Santa Fe de Guanajuato y la Gestión del Ing. Alfredo Terrazas Vega (1947–1972). Guanajuato: University of Guanajuato.
1996 Mineral de la Luz. Guanajuato: Instituto de la Cultura del Estado de Guanajuato.
2002 Ponciano Aguilar y su Circunstancia. Guanajuato: Ediciones la Rana, Instituto de la Cultura del Estado de Guanajuato.

Jiménez, Francisco
1857 Diario-Memoria de los trabajos científicos practicados bajo la dirección de Francisco Jiménez, Ingeniero de la Comisión de Límites Mexicana conforme a las instrucciones del Señor Como Don José Salazar Ylarregui, a quien se hace entrega de ellos. Unpublished MS. Special Collections, University Library, University of Texas at El Paso.

Johnson, Paul W.
1965 A Field Guide to the Gems and Minerals of Mexico. New York: Gembooks.

Jones, Bob
2004 A Fifty-Year History of the Tucson Show. Tucson: Mineralogical Record.

Kampf, Anthony
2006 The Ben Frankenberg Collection of Bisbee Minerals. Mineralogical Record 37(4):275–282.

Kearney, Michael
1995 The Local and the Global: The Anthropology of Globalization and Transnationalism. Annual Review of Anthropology 24:547–565.

Knight, Alan
1990 The Mexican Revolution: vol. 2. Counter-Revolution and Reconstruction. Lincoln: University of Nebraska Press.

Kohlstedt, Sally Gregory
1980 Henry A. Ward: The Merchant Naturalist and American Museum Development. Journal of the Society for the Bibliography of Natural History 9:647–661.

Kohout, Martin Donell
N.d. Bartlett-Garcia Conde Compromise. Handbook of Texas Online. http://www.tshaonline.org/handbook/online/articles/nbb02, accessed May 12, 2011.

Kopytoff, Igor
1986 The Cultural Biography of Things: Commoditization as Process. In In The Social Life of Things: Commodities in Cultural Perspective. Arjun Appadurai, ed. Pp. 64–91. Cambridge: Cambridge University Press.

Kunz, George
1927 American Travels of a Gem Collector. Saturday Evening Post, November 26, pp. 6–7, 85–86, 91; December 10, 1927, pp. 22–23, 172–174, 176.

Latour, Bruno
1988 Science in Action: How to Follow Scientists and Engineers through Society. Cambridge: Harvard University Press.

2005   Reassembling the Social: An Introduction to Actor-Network-Theory. Clarendon Lectures in Management Studies. New York: Oxford University Press.

La Venta
2007   http://naica.laventa.it/naica-expedition-07.en.html, accessed May 1, 2008.

León, Luis D.
2004   La Llorona's Children: Religion, Life and Death in the US-Mexican Borderlands. Berkeley: University of California Press.

Lefebvre, Henri
1991 [1968]   The Production of Space. Donald Nicholson-Smith, trans. Oxford: Blackwell Press.

Lévi-Strauss, Claude
1992 [1955]   Tristes Tropiques. New York: Penguin Books.

Lewis, Laura A.
2006   Home Is Where the Heart Is: Afro-Latino Migration and Cinder-Block Homes on Mexico's Costa Chica. South Atlantic Quarterly 105(4):801–829.

Logan, Michael F.
2006   Desert Cities: The Environmental History of Phoenix and Tucson. Pittsburgh: University of Pittsburgh Press.

London, David
2003   New "Cave of the Crystals" at Naica, Chihuahua, Mexico. Earth Scientist Magazine 2003:24–27. https://cq5publish.ou.edu/content/dam/mcee/geology/earth-scientist/EarthScientist2003.pdf, accessed.

Lowe, Celia
2004   Making the Monkey: How the Togean Macaque Went from "New Form" to "Endemic Species" in Indonesians' Conservation Biology. Cultural Anthropology 19(4):491–516.

Lutz, Catherine, and Jane Collins, eds.
1993   Reading National Geographic. Chicago: University of Chicago Press.

Magazine, Roger
2007   Golden and Blue like My Heart: Masculinity, Youth and Power among Soccer Fans in Mexico City. Tucson: University of Arizona Press.

Malinowski, Bronislaw
1984 [1922]   Argonauts of the Western Pacific. Long Grove, IL: Waveland Press.

Marmolejo, Padre Lucio
1988 [1886]   Efemerídes Guanajuatenses. Guanajuato: Estado de Guanajuato.

Martin, Emily
1995   Working across the Human-Other Divide. In Reinventing Biology: Respect for Life and the Creation of Knowledge. Lynda Birke and Ruth Hubbard, eds. Pp. 261–275. Bloomington: Indiana University Press.

Martin, Percy F.
1905   Mexico's Treasure House: Guanajuato. New York: Cheltenham Press.

Martínez, Torres, Lucila, Elia Mónica Morales Zárate, and Maria Jesús Puy y Alquiza
2006   Museo de Mineralogía 1870–1972 Eduardo Villaseñor Söhle. Facultad de Minas, Metalurgía y Geología, Universidad de Guanajuato. Boletín de Mineralogía 17:173–174.

Massey, Doreen.
    1991   A Global Sense of Place. Marxism Today, June:24–29.
Matsutake Worlds Research Group
    2009   A New Form of Collaboration in Cultural Anthropology: Matsutake Worlds.
    American Ethnologist 36(2):380–403.
Mauss, Marcel
    1990 [1950]   The Gift: The Form and Reason for Exchange in Archaic Societies.
    New York: W. W. Norton.
Meieran, Eugene
    2008   Introduction. *In* The Miguel Romero Collection of Mexican Minerals. T. C.
    Wallace, W. E. Wilson, P. K. M. Megaw, and R. C. Currier. Pp. 3–4. Tucson:
    Mineralogical Record.
Mendoza, Cristóbal
    2006   Transnational Spaces through Local Places: Mexican Immigrants in Albu-
    querque, New Mexico. Journal of Anthropological Research 62(4):539–561.
Meneley, Anne
    2008   Oleo-Signs and Quali-Signs: The Qualities of Olive Oil. Ethnos 73(3):303–
    326.
Metropolis, William C.
    1999   Highlights of Mexican Minerals at Harvard University. Rocks & Minerals
    74(1):56–61.
Meyer Cosío, Francisco Javier
    1999   La Minería en Guanajuato: Denuncios, Minas, y Empresas (1898–1913).
    Zamora, Michoacán: El Colegio de Michoacán / Universidad de Guanajuato.
Milenio
    2010   Naica y sus cristales, en las Rejas de Chapultepec, July 7.
Miller, Daniel
    1987   Material Culture and Mass Consumption. Oxford: Blackwell Press.
    2005   Materiality: An Introduction. *In* Materiality. D. Miller, ed. Pp. 1–50.
    Durham: Duke University Press.
    2008   The Uses of Value. Geoforum 39:1122–1132.
Mineralogical Magazine
    1914   Obituary of Albert Fairchild Holden. Mineralogical Magazine 17(79):117–118.
Mineralogical Record
    1998   Mexico—I. Special issue, Mineralogical Record 29(1).
    2003a   Mexico—II. Special issue, Mineralogical Record 34(5).
    2003b   Mexico—III. Special issue, Mineralogical Record 34(6).
    2004   Mexico—IV. Special issue, Mineralogical Record 35(6).
Mintz, Sidney
    1985   Sweetness and Power: The Place of Sugar in Modern History. New York:
    Viking Press.
Misra, Kura
    2000   Understanding Mineral Deposits. New York: Springer Press.
Mitchell, Don
    1996   The Lie of the Land: Immigrant Labor and the California Landscape. Min-
    neapolis: University of Minnesota Press.
Mitchell, Timothy
    1991   Colonizing Egypt. Berkeley: University of California Press.

Miyazaki, Hiro

    2003   The Temporalities of the Market. American Anthropologist 105(2):255–265.

Mol, Annemarie

    2002   The Body Multiple: Ontology in Medical Practice. Durham: Duke University Press.

Molella, Arthur P.

    1991   The Museum That Might Have Been: The Smithsonian's National Museum of Engineering and Industry. Technology and Culture 32(2):237–263.

Moore, T. P., and Marcus J. Origlieri

    2008   Famous Mineral Localities: The Milpillas Mine, Cananea District, Sonora, Mexico. Special issue, "Mexico-V." Mineralogical Record 39(Nov–Dec):25–34.

Moore, T. P., and P. K. M. Megaw

    2003   Famous Mineral Localities: The Ojuela Mine, Mapimí, Durango, Mexico. Mineralogical Record 34(5):1–91.

Morales Zárate, Elia Mónica, Lucila Martínez Torres, and María Guadalupe Villaseñor Cabral

    2008   Museo de Mineralogía "Eduardo Villaseñor Söhle." N.p.: Goldcorp Mexico.

Morelos Rodríguez, Lucero

    2010   Ciencia, Estado, y Científicos: El Desarrollo de la Geología Mexicana A Traves del Estudio de los Ingenieros Antonio del Castillo, Santiago Ramírez y Mariano Barcena (1843–1902). Masters' thesis, Department of History, Universidad Nacional Autónoma de México.

Mueggler, Erik

    2011   The Paper Road: Archive and Experience in the Botanical Exploration of West China and Tibet. Berkeley: University of California Press.

Munn, Nancy

    1977   The Spatiotemporal Transformations of Gawa Canoes. Societé des Océanistes 33(54–55):39–53.

    1986   The Fame of Gawa: A Symbolic Study of Value Transformation in a Massim [Papua New Guinea] Society. Cambridge: Cambridge University Press.

Myers, Fred

    2002   Painting Culture: The Making of an Aboriginal High Art. Durham: Duke University Press.

Myers, Fred, ed.

    2002   The Empire of Things: Regimes of Value and Material Culture. Santa Fe: School of American Research Press.

Naica

    2012   http://www.slideshare.net/random13579/naica-crystal-cave, accessed October 22, 2012.

Nash, June

    1979   We Eat the Mines and the Mines Eat Us: Dependency and Exploitation in Bolivian Tin Mines. New York: Columbia University Press.

National Geographic Channel

    2010   Giant Crystal Cave. (documentary)

Nature: A Weekly Illustrated Journal of Science

    1891   May 28: 89.

New, Mike

    2003   Mike New's Story, in Purple Adamite Tales [sidebar]. Mineralogical Record 34(5):44–52.

Newell, Julie R.

    1997   James Dwight Dana and the Emergence of Professional Geology in the United States. Special issue, American Journal of Science 297(3):273–282.

Nickel, E. H., and J. D. Grice

    1998   The IMA Commission on New Minerals and Mineral Names: Procedures and Guidelines on Mineral Nomenclature. Canadian Mineralogist 36:913–926.

Norget, Kristin

    2006   Days of Death, Days of Life: Ritual in the Popular Culture of Oaxaca. New York: Columbia University Press.

Palerm, Angel

    1980   El Primer Sistema Económico Mundial. *In* Antropología y Marxismo. Mexico City: Editorial Nueva Imagen.

Panczner, William D.

    1987   Minerals of Mexico. New York: Van Nostrand Reinhold Company.

Paxson, Heather

    2008   Post-Pasteurian Cultures: The Microbiopolitics of Raw-Milk Cheese in the US. Cultural Anthropology 23(1):15–47.

Paz, María Emilia

    1997   Strategy, Security and Spies: Mexico and Its Allies in World War II. University Park: Pennsylvania State University.

Petruk, W., and D. Owens

    1974   Some Mineralogical Characteristics of the Silver Deposits of the Guanajuato Mining District, Mexico. Economic Geology 69(7):1078–1085.

Petruk, W., D. R. Owens, A. J. Stewart, and E. G. Murray

    1974   Observations of Acanthite, Aguilarite and Naumannite. Canadian Mineralogist 12:365–369.

Polanyi, Karl

    1944   The Great Transformation: The Political and Economic Origins of Our Time. Boston: Beacon Press.

Pomian, Krzysztof

    1990   Collectors and Curiosities: Paris and Venice, 1500–1800. Cambridge: Polity Press.

Price, Sally

    2007   Paris Primitive: Jacques Chirac's Museum on the Quai Branly. Chicago: University of Chicago Press.

Public Broadcasting Service

    2009   American Experience: Murder at Harvard.

Ramírez, Santiago E.

    1890   Datos para la historia del Colegio de Minería. Mexico City: Edición de la Sociedad "Alzate."

    1891   Biografía del sr. D. Andrés Manuel del Río: Primer Catedrático de Mineralogía del Colegio de Minería. México: Imprenta del Sagrado Corazón de Jesús.

Rankine, Margaret E.

    1992   The Mexican Mining Industry in the Nineteenth Century with Special Reference to Guanajuato. Bulletin of Latin American Research 11(1):29–48.

Raymer, Robert G.
  1935 Early Copper Mining in Arizona. Pacific Historical Review 4(2):123–130.
Rebert, Paula
  2001 La Gran Línea: Mapping the United States–Mexico Boundary, 1849–1857.
    Austin: University of Texas Press.
Redfield, Robert
  1950 The Folk Culture of Yucatán. Chicago: University of Chicago Press.
Reuters
  2008 February 9.
Rickard, T. A.
  1907 Journeys of Observation. San Francisco: Dewey Publishing Company.
Rikowski, Glenn
  2008 Forms of Capital: Critique of Bourdieu on Social Capital. The Flow of
    Ideas. http://www.flowideas.co.uk/index.php, accessed April 15, 2008.
Robb Report
  2002 Rocks of Ages. September 21.
Rockhounds Forum
  2002 July, www.rockhounds.com, accessed August 15, 2004.
Rogers, Douglas
  2005 Moonshine, Money, and the Politics of Liquidity in Rural Russia. American
    Ethnologist 32(1):63–81.
Rosenblum, Jonathan
  1995 Copper Crucible: How the Arizona Miner's Stike of 1983 Recast Labor-
    Management Relations in America. Ithaca, NY: ILR Press.
Ross, Brad, Sean Dessurealt, and Michael Rieber
  2010 The Tucson Mineral Show and the Market for Collector Minerals:
    The Potential for Artisanal and Small-Scale Miners. Resources Policy
    36:168–177.
Rouse, Roger
  1991 Mexican Migration and the Social Space of Postmodernism. Diaspora
    1(1):8–23.
Rubinovich Kogan, Raúl
  1993 Desarrollo histórico de la mineralogía en México. Primer Congreso Mexi-
    cano de Mineralogía, Memoria 1993:184–189.
Rydell, Robert
  1987 All the World's a Fair: Visions of Empire at American International Exposi-
    tions, 1876–1916. Chicago: University of Chicago Press.
Sahlins, Marshall
  1988 Cosmologies of Capitalism: The Trans-Pacific Sector of "the World System."
    Proceedings of the British Academy 74:1–51.
Salazar Ylarregui, José
  1851 Datos de los trabajos astronómicos y topográficos dispuestos en forma de
    diario. Mexico City: Imprenta de Juan N. Navarro.
  N.d. Archivo Histórico Genaro Estrada, folder 40-16-139.
Saussure, Ferdinand de
  1959 [1916] General Course in Linguistics. New York: Philosophical Press.
Schama, Simon
  1991 Dead Certainties: Unwarranted Speculations. New York: Vintage Books.

Scott, Joan Wallach
  1989   Gender and the Politics of History. New York: Columbia University Press.
Silver Institute
  N.d.   Historical COMEX spot prices. http://www.silverinstitute.org/site, accessed
  June 1, 2010. journal
Skinner, H. C. W.
  2005   Biominerals. Mineralogical Magazine 69(5):621–641.
Smithsonian Institution
  1856   Smithsonian Annual Report. Washington, DC: Smithsonian Institution
  Press.
Smithsonian Institution Archives
  Accession 93-121, Paul E. Desautels papers. Washington, DC.
  Record Unit 7281, William F. Foshag Collection. Washington, DC.
  Record Unit 7002, Spencer F. Baird Papers. Washington, DC.
Smithsonian Institution Libraries
  2012   The Will of James Smithson. http://www.sil.si.edu/Exhibitions/Smithson
  -to-Smithsonian/will.htm, accessed October 24, 2012.
Soto Laveaga, Gabriela
  2009   Jungle Laboratories: Mexican Peasants, National Projects, and the Making
  of the Pill. Durham: Duke University Press.
Stephen, Lynn
  2007   Transborder Lives: Indigenous Oaxacans in Mexico, California, and Oregon.
  Durham: Duke University Press.
Stewart, Susan
  1993   On Longing: Narratives of the Miniature, the Gigantic, the Souvenir, the
  Collection. Durham: Duke University Press.
Steiner, Christopher B.
  1995   The Art of the Trade: On the Creation of Value and Authenticity in the African Art Market. *In* The Traffic in Culture: Refiguring Art and Anthropology.
  George E. Marcus and Fred R. Myers, eds. Pp. 151–165. Berkeley: University of
  California Press.
Strathern, Marilyn
  1990   The Gender of the Gift: Problems with Women and Problems with Society
  in Melanesia. Berkeley: University of California Press.
Tamayo, P. de Ham, Luz María Oralia, and José Omar Moncada Maya
  2001   La Comisión de Límites de México y el levantamiento de la línea divisoria
  entre México y Estados Unidos, 1849–1857. Investigaciones Geográficas, Boletín
  del Instituto de Geografía, UNAM 44:85–102.
Taussig, Michael
  1982   The Devil and Commodity Fetishism in South America. Chapel Hill: University of North Carolina Press.
Thomas, Nicholas
  1991   Entangled Objects: Exchange, Material Culture, and Colonialism in the Pacific. Cambridge: Harvard University Press.
  1994   Licensed Curiosity: Cook's Pacific Voyages. *In* The Cultures of Collecting.
  John Elsner and Roger Cardinal, eds. Pp. 116–136. Cambridge: Harvard University Press.

Thompson, E. P.
  1971 The Moral Economy of the English Crowd in the Eighteenth Century. Past and Present 50:76–136.
Thompson, Wayne
  2007 Ikons, Classics, and Contemporary Masterpieces. Tucson: Mineralogical Record.
Time
  1944 El Monstruo. Time, October 2.
Tolentino, Roland B.
  1996 Bodies, Letters, Catalogs: Filipinas in Transnational Space. Social Text 48:49–76.
Trouillot, Michel-Rolph
  1991 Anthropology and the Savage Slot: The Poetics and Politics of Otherness. *In* Recapturing Anthropology. Richard G. Fox, ed. Pp. 17–44. Santa Fe: School of American Research Press.
  1994 Silencing the Past: Power and the Production of History. Boston: Beacon Press.
Tsing, Anna
  2001 The Global Situation. Cultural Anthropology 15(3):327–360.
  2009 Beyond Economic and Ecological Standardization: Supply Chains and the Human Condition. Australian Journal of Anthropology 20(3):347–368.
Turner, Terence
  1995 Social Body and Embodied Subject: Bodiliness, Subjectivity and Sociality among the Kayapó. Cultural Anthropology 10(2):140–170.
Tutino, John
  2011 Making a New World: Founding Capitalism in the Bajío and Spanish North America. Durham: Duke University Press.
United States Army Corps of Topographical Engineers and J. D. Graham
  1799–1865 Report of the Secretary of War. August 31, 1852. http://scholarship.rice.edu/jsp/xml/1911/26932/1/aa00367.tei.html#div2194,%20ram%EDrez, accessed April 14, 2009.
United States Department of the Interior
  1996 Dictionary of Mining, Mineral, and Related Terms. 2nd edition. Staff of the U.S. Bureau of Mines, comp. and ed. http://xmlwords.infomine.com/xmlwords.htm, accessed June 10, 2010.
Uribe Salas, José Alfredo
  2006 Labor de Andrés Manuel del Rio en México: Profesor en el real seminario de minería e innovador tecnológico en minas y ferrerías. Asclepio. Revista de Historia de la Medicina y de la Ciencia 58(2):231–260.
Uribe Salas, José Alfredo, and María Teresa Cortés Zavala
  2006 Andrés del Río, Antonio del Castillo y José G. Aguilera en el desarrollo de la Ciencia Mexicana. Revista de Indias 66(237):491–518.
Vassallo, Luis Fernando
  1988 Características de la Composición Mineralógica de las Menas de la Veta Madre de Guanajuato. Universidad Nacional Autónoma México, Institución Geología Revista 7(2):232–243.

Vassallo, Luis Fernando, and Margarita Reyes-Salas
  2007   Selenium Polybasite from the Guanajuato Mining District, Mexico. Bol-E
      3(2):1–16.
Villanueva, Hecky
  2007   Desert Pirates on the Ultimate Treasure Hunt? The Annual Tucson Gem,
      Mineral, and Fossil Showcase. Unpublished MS. In possession of E. E. Ferry,
      Brandeis University, Waltham, MA.
Villarreal Garza, J. Trinidad, Luis Lauro Flores Flores, and Pedro Castro Acevedo
  1998   Programa de Aparcería Tyson de México, S.A. de C.V. Revista Mexicana de
      Agronegocios 3(July–December).
Wallace, Terry
  2008   Miguel Romero Sánchez (1924–1997). *In* The Miguel Romero Collection
      of Mexican Minerals. T. C. Wallace, W. E. Wilson, P. K. M. Megaw, and R.C.
      Currier. Pp. 5–10. Tucson: Mineralogical Record.
Walsh, Andrew
  2010   The Commodification of Fetishes: Telling the Difference between Natural
      and Synthetic Sapphires. American Ethnologist 37(1):98–114.
  2012   Made in Madagascar: Sapphires, Ecotourism, and the Global Bazaar.
      Toronto: University of Toronto Press.
Ward, Henry G.
  1828   Mexico in 1827. London: Henry Colburn.
Webmineral
  N.d.   Alphabetical Listing of Mineral Species. http://webmineral.com/data
      /Aguilarite.shtml, accessed May 12, 2008.
Weiner, Annette B.
  1988   Women of Value, Men of Renown: New Perspectives in Trobriand Exchange.
      Austin: University of Texas Press.
  1992   Inalienable Possessions: The Paradox of Keeping-while-Giving. Berkeley:
      University of California Press.
Werne, Joseph Richard
  1987   Major Emory and Captain Jiménez: Running the Gadsden Line. Journal of
      the Southwest 29(2):203–222.
Wilensky, Stuart
  2009   About Wilensky Fine Minerals. http://www.wilenskyminerals.com/about/,
      accessed October 12, 2009.
Wilson, Wendell
  N.d.   Guide to Writing Locality Articles. Mineralogical Record. http://minrec
      .org/article.asp, accessed January 3, 2010.
  1990   Connoisseurship in Minerals. Mineralogical Record 21(1):7–12.
  1994   A History of Mineral Collecting. Mineralogical Record 25(6).
  2006   Early Mineral Dealers: William Niven (1850–1937). Mineralogical Record
      37(4):297–309.
  2008   The Miguel Romero Collection. *In* The Miguel Romero Collection of Mexi-
      can Minerals. T. C. Wallace, W. E. Wilson, P. K. M. Megaw, and R. C. Currier.
      P. 11. Tucson: Mineralogical Record.
  2012   Mineralogical Record Biographical Archive. http://www.mineralogicalre-
      cord.com, accessed February 2, 2012.

Wood, W. Warner
  2008   Made in Mexico: Zapotec Weavers and the Ethnic Art Market. Bloomington: Indiana University Press.
Zelizer, Viviana A.
  1994   Pricing the Priceless Child: The Changing Social Value of Children. Princeton: Princeton University Press.
Zlolniski, Christian
  2006   Janitors, Street Vendors, and Activists: The Lives of Mexican Immigrants in Silicon Valley. Berkeley: University of California Press.

INDEX

Harvard mineralogical collections;
Harvard Mineralogical Museum;
Harvard Museum of Natural History
Haüy, René Just, 56, 70
Hayden, Corinne, 14–15
Helms, Mary, 159–60
Hill, Jane, 173, 207n8
Holden, Albert F., 72, 122, 124, 196
horizon, in minerals, 152, 159
Houston Museum of Natural Science,
78, 135
Hunt, T. Sterry, 37
Hutton, James, 202n1
hybridity, 12, 14

imperialism, 9, 27, 114, 124
inalienability, 127, 132, 201n2
inequality, 13, 24, 93, 105–106, 116–17, 162
Ingold, Tim, 11–12
Instituto de Geología, Instituto Geológico de México (Universidad
Nacional Autónoma de México),
64, 66, 68
Internet, 8, 28, 39, 74, 171, 187

Jiménez, Francisco, 105

kinship, 6, 144
Kothavala, Rustam Z., 76, 131–32
Kunz, George F., 3, 7, 122

La Aurora mine, 54
La Luz, 30, 32, 91, 123, 149
labor, 11, 13, 16–18, 31, 37, 50, 82, 142–43,
145–48, 150–51, 153, 155, 163–64, 166,
169, 175–76, 182, 192, 201n2, 207n1
laboratories, scientific, 9, 118, 124, 153
Landero y Cos, José, 73
lapidarists, 73–74, 152, 170
Latin America, 30, 130, 140, 147
Latour, Bruno, 21–22
Lavoisier, Antoine-Laurent de, 56, 58
lead, 6, 41, 44–45, 47–48, 66, 106, 178–
79, 182, 202nn2,3 (chap2)
Lefebvre, Henri, 11–12
legrandite, 6, 41, 46, 48–50, 54–55, 126,
163, *164*, 165, 176–77, 186
Lévi-Strauss, Claude, 15

Los Angeles County Museum, 14, 37, 78
Lyell, Charles, 202n1

malachite, 3, 37, 189–90, 202n4
Mapimí, 3, 5–7, 9, 25, 27, 32, 43–50, 55,
66, 126, 143, 158, 163–67, 175–81, *183*,
184, 186–92, 208n13, 208n21
mapimite, 126
marketplaces, 11, 27, 28, 42, 113, 137, 163,
166–67, 170, 173, 192
markets, 1, 7–9, 15, 20, 23, 25, 27, 40,
48–49, 70, 72, 76, 78, 80–81, 135–36,
137, 144, 152, 154, 163–68, 175–77,
180, 185–92, 201n4, 203n1, 205n7,
207n8 (chap5), 207n1, 207n7 (chap6),
209n22; global, 16, 30; market exchange, 127, 135
Marx, Karl, 16–17, 154
masculinity, 24, 141, 174, 176, 181
Massey, Doreen, 14
mendozavilite, 88
Meneley, Anne, 10
Menger, Carl, 16
Mexican Revolution, 44, 61, 64, 150
Mexico City, 6, 31, 43, 57–58, 63–65, 68,
73, 111, 118
middlemen, 49, 175–77, 186–87, 190. *See
also coyotes* (middlemen)
migration, 10, 12–13, 16, 37, 142; Mexico–
United States, 13, 31, 180–81
miguelromeroite, 203n3
Miller, Daniel, 17–18
Mineralogical Record, 38, 81–82, 92, *125,*
133–36, 156, 158–59, 167–68
mineralogy, 3, 6–8, 10, 15, 26, 28, 47, 52,
56–58, 60–64, 67–68, 70, 73, 75–76,
88, 91–92, 94–95, 99, 102, 120–22, 129,
131–33, 202n10, 205n6
miners, 5–7, 9, 12, 24, 27, 28, 33–34, 37,
43–44, 46–50, 75, 82, 96, 106, 111, 118,
124, 127, 134, 136, 137, 139–41, 143–50,
153, 159–61, 165, 167–69, 176–82, 184–
85, 188–89, 191–92, 194, 197, 203n1,
206n3, 208nn13,14, 208n16; drillers,
33, 143, 169. *See also* mines; mining
mines, 1, 6–9, 11, 36, 39, 41, 60–61,
72–73, 83, 88, 99–100, 113, 121–24,
137, 145–49, 152–53, 159, 178, 182, 184,

TRACKING GLOBALIZATION

Illicit Flows and Criminal Things: States, Borders,
and the Other Side of Globalization
EDITED BY WILLEM VAN SCHENDEL AND ITTY ABRAHAM

Globalizing Tobacco Control: Anti-Smoking
Campaigns in California, France, and Japan
RODDEY REID

Generations and Globalization: Youth, Age,
and Family in the New World Economy
EDITED BY JENNIFER COLE AND DEBORAH DURHAM

Youth and the City in the Global South
KAREN HANSEN IN COLLABORATION WITH ANNE LINE DALSGAARD,
KATHERINE GOUGH, ULLA AMBROSIUS MADSEN,
KAREN VALENTIN, AND NORBERT WILDERMUTH

Made in Mexico: Zapotec Weavers and the Global Ethnic Art Market
W. WARNER WOOD

The American War in Contemporary Vietnam:
Transnational Remembrance and Representation
CHRISTINA SCHWENKEL

Street Dreams and Hip Hop Barbershops:
Global Fantasy in Urban Tanzania
BRAD WEISS

Aging and the Indian Diaspora:
Cosmopolitan Families in India and Abroad
SARAH LAMB

Cultural Critique and the Global Corporation
EDITED BY PURNIMA BOSE AND LAURA E. LYONS

Recycling Indian Clothing: Global Contexts of Reuse and Value
LUCY NORRIS

Music and Globalization: Critical Encounters
EDITED BY BOB W. WHITE

The Virtual Village: Filipino Migrants Coping with a Global World
DEIRDRE MCKAY

Minerals, Collecting, and Value across the U.S.-Mexico Border
ELIZABETH EMMA FERRY

ELIZABETH EMMA FERRY is an associate professor of Anthropology at Brandeis University. She has been studying silver mining in the central Mexican city of Guanajuato since 1994. Her books include *Not Ours Alone: Patrimony, Value and Collectivity in Contemporary Mexico* and *Timely Assets: The Politics of Resources and their Temporalities* (coedited with Mandana Limbert).

CPSIA information can be obtained at www.ICGtesting.com
Printed in the USA
LVOW132315100613

337914LV00003B/3/P